Window into the Jurassic World

Dilophosaurus

Window into the Jurassic World

Dinosaur State Park, Rocky Hill, Connecticut

Nicholas G. McDonald

With Photography by Richard Bergen

Friends of Dinosaur State Park and Arboretum, Inc.
Rocky Hill, Connecticut

Published by the Friends of Dinosaur Park and Arboretum, Inc.
400 West Street, Rocky Hill, CT
06067

Printed in the United States of America by Lithographics, Inc., Farmington, CT

This book is printed with vegetable-based inks on Chorus Art Gloss paper
that is FSC certified and contains 30% recycled content.

Photographs by Bergen © 2010 Richard Bergen Photography, Hartford, CT

Graphic Design by John Alves, Hartford, CT

ISBN: 978-0-9825905-0-8 (hardcover)

ISBN: 978-0-9825905-1-5 (softcover)

Copies of this book may be purchased at the FDPA Bookshop at Dinosaur State Park
www.dinosaurstatepark.org

Front Cover: The life-size model of *Dilophosaurus* at Dinosaur State Park.
Rear Cover: A 16-inch-long *Eubrontes* footprint, Dinosaur State Park.

CONTENTS

Preface

When asked where dinosaur fossils are found, people tend to say, "the Western United States," "Mongolia," or other exotic locales. Even experts are surprised to learn that the first documented dinosaur bones in the Western Hemisphere were found in Connecticut, the first dinosaur footprints described in the world came from Massachusetts, and the largest collection of dinosaur tracks anywhere is from these two States. In fact, Connecticut and Massachusetts are home to some of the world's richest fossil deposits dating from early in the Age of Dinosaurs. In addition to the bones and tracks of dinosaurs, Jurassic-age rocks in New England contain exquisitely preserved fishes, as well as plant and invertebrate fossils. These fossils are found within the Central Valley, a rift basin that formed as North America and Africa split apart during the Mesozoic Era.

Perhaps the most notable of the fossil discoveries in the Central Valley are the remarkably abundant dinosaur footprints preserved in place at Dinosaur State Park in Rocky Hill, Connecticut. *Window into the Jurassic World* is the story of the Park's trackways and other Mesozoic fossils, as told by Nicholas G. McDonald, whose passion for research into the life of the past spans more than four decades. Regarded as the foremost authority on the region's paleontology, he has authored or coauthored numerous scholarly papers and abstracts, and has published a bibliographic guidebook to the geologic literature on the Valley. In this new book, Nick takes us from concepts of geological time and plate tectonics, through the geologic history and paleontology of the Valley, up to the present day, emphasizing how modern landscapes reflect their ancient bedrock roots. This book is not only a guide to the fossil exhibits, "Arboretum of Evolution" and ecological habitats at the Park, it is also a concise portrayal of the natural history of Connecticut and Massachusetts at the beginning of the Age of Dinosaurs.

The paleontological and geological history of the Central Valley has significance extending far beyond the local region. First, the sedimentary rocks of the Valley are of Late Triassic and Early Jurassic age. These rocks preserve a uniquely detailed record of the aftermath of one of the largest mass extinctions of all time, that which occurred at the Triassic-Jurassic boundary, about 201 million years ago. The dinosaur trackways at the Park are typical of what is seen globally following this mass extinction, reflecting what appear to be ecologically odd communities in which carnivores seem to be more abundant than herbivores. Second, the fossil-rich sedimentary strata are interbedded with extensive basaltic lava flows, exposed as the familiar "traprock" ridges of the Valley. The rugged, igneous rock cliffs and ridges that are a prominent feature of central Connecticut and Massachusetts are also part of the most widespread lava eruptions on Earth. Now deeply eroded, early Mesozoic lava flows originally covered as much as four million square miles of the Earth's surface, an area nearly one-third as large as the surface of the Moon! These massive eruptions may have been the cause of the Triassic-Jurassic boundary extinctions. Third, much of the sedimentary record in the Central Valley formed in large, rift valley lakes similar to those found in East Africa today. These great "fossil lakes" were populated by myriad species of fishes that have been preserved in layers of black shale, along with other aquatic animals and plants. Lake levels rose and fell not only due to seasonal weather patterns, but also because of changes in climate, caused by cyclical variations in the Earth's orbit. The Central Valley's Early Jurassic lake deposits provide some of the best evidence of orbitally paced climate changes yet obtained.

Window into the Jurassic World provides the most detailed account to date of the discovery and preservation of the trackways at Dinosaur State Park, and describes their geological and ecological context during the Mesozoic Era. The book is an authoritative and accessible narrative, further distinguished by having outstanding graphics and photographs on nearly every page. Some of the finest fossils obtained from the Valley are illustrated herein for the first time. This volume is an ideal companion to Park visits, an informative guide to exceptional local fossils and their Mesozoic environments, and a superb entrée into the subjects of paleontology and geology in general.

Paul E. Olsen
Storke Memorial Professor of Earth and Environmental Sciences,
Columbia University, New York, NY

Acknowledgments

The "seeds" for the writing of this book were planted many years ago by Richard Krueger, a friend and geological colleague, and the director of Dinosaur State Park for more than three decades. Rich always made me feel welcome at the Park, where we discussed fossils and his countless Park projects and plans. Rich convinced me that the Park would be an ideal place for the public to view some of my best Central Valley fossils. In 1996, after the Exhibit Center was renovated, a number of fishes, plants and other Jurassic fossils from my collection were put on display at the Park in specially designed cases. As a logical next step, Rich and others then advocated that someone should produce a publication on the Park's history, trackways and exhibits.

Encouragement to finally put pen to paper on a Dinosaur Park book came from Chris Sprague, Jan Locke and Meg Enkler. Rich Krueger, now retired, offered his services as a "tour guide" through the Arboretum and Nature Trails for photographer Richard Bergen and me, and he generously shared his unique knowledge of the natural areas he had conserved and developed. Rich wrote initial drafts for the Arboretum and Nature Trails chapters, created the silhouette drawings for the Arboretum plantings, and identified plant species and cultivars.

It has been a distinct pleasure working with Richard Bergen for four years on this project. He is a consummate professional, with an eye for composition, lighting, aesthetics and detail. In spite of long hours in the field, calls for additional shots and endless photo editing, Richard's enthusiasm for the project has never wavered, nor has his amiable manner and good humor. I am grateful for his support and friendship.

Discussions with Paul Olsen and Emma Rainforth on ichnological topics were most beneficial. Rich Krueger, Morgan Shipway, Meg Enkler, Christine Witkowski, Ralph Lewis and members of the "Friends" Board provided valuable reviews of the first draft of the manuscript. I am particularly indebted to Meg and Christine of the Park staff for their role as copy editors of subsequent drafts. Their constructive critiques and excellent wordsmithing greatly enhanced the clarity and flow of the text, and their steady encouragement helped me maintain my own enthusiasm. Morgan Shipway and David Wakefield proofread the final draft.

Paul Olsen and Christine Witkowski drafted and provided figures in the text and assisted in obtaining some of the photographs. Tekla Harms and Kate Wellspring provided permission and access to photograph specimens at the Amherst College Museum of Natural History. Jacques Gauthier, Tim White and Cope MacClintock allowed access to specimens at the Yale Peabody Museum of Natural History. I am grateful to the Friends of Dinosaur Park and Arboretum, Inc. for providing the opportunity to write this book, for their financial assistance and for their guidance throughout this project. Last, and certainly not least, I thank my wife, Pam, for help with hours of proofreading, and for her abiding, unflagging support, which has made this endeavor a labor of love.

DEDICATION

This book is dedicated to former Park director Richard Krueger, who spent more than thirty years transforming Dinosaur State Park into a unique educational facility. A talented geologist, botanist and artist, Rich developed indoor and outdoor exhibits, designed and constructed nature trails, and created the splendid Arboretum of ancient species that now surrounds the Exhibit Center. Under his guidance, a track-filled construction site evolved into a world-class museum and nature sanctuary that will provide instructional and inspirational experiences far into the future. Thank you, Rich.

This book is also dedicated to the memory of Chris Sprague, Dinosaur State Park's first official volunteer, who worked tirelessly for two decades to promote the Park. Her unwavering efforts, enthusiasm and encouragement brought the staff and volunteers together as a family that worked side by side to turn vision into reality. Many of the exhibit features and outdoor gardens at the Park are examples of that shared concern. Chris was a true Friend.

Chapter 1

Introduction

August 23, 1966, was an important day for Science.

It was a day that would broaden the understanding and appreciation of Planet Earth by many of its human residents. On that date, Lunar Orbiter I transmitted the first picture of Earth taken by a spacecraft orbiting the Moon. That photograph showed a drab, lifeless, cratered lunar surface in the foreground and a bright, dynamic, cloud-draped Earth in the background. It was the first of many satellite photos (Figure 1-1) that would inspire a view of the Earth not just as a collection of peoples or countries, but rather as a constantly changing, fragile but resilient "living planet" that has nurtured life for more than two billion years.

That same day, a bulldozer operator digging a foundation in Rocky Hill, Connecticut, pulled up a large slab of bedrock. His attention was drawn to unusual markings on the underside of the sandstone block. Rather than continuing his excavations, he paused to examine the markings, and brought them to the attention of his co-workers. The markings would prove to be the well-preserved footprints of *Jurassic*-age

Figure 1-1. As their spacecraft rocketed toward the Moon on December 7, 1972, the Apollo 17 astronauts snapped this now-famous photograph of the fully sunlit Earth, 18,000 miles below. Much of the coastline of Africa is visible, as is the frozen Antarctic continent. This and similar images from space have inspired greater consideration of the collective role of all people as caretakers of our unique planet.

dinosaurs, and the site was soon proclaimed to contain one of the largest concentrations of dinosaur trackways in the world. That construction site is now known as Dinosaur State Park (Figure 1-2). Since the discovery of the tracks, more than two million Park visitors have gained a greater understanding of Earth's distant past and the plants and animals that flourished during ancient times, now preserved as *fossils*.

Ever since the discovery and identification of fossilized dinosaur bones in the early decades of the 19th century, humans have been fascinated by these strange, often massive, ancient residents of our planet. The name "dinosaur," first introduced at an 1841 meeting of scientists by British anatomist Richard Owen, translates to "terrible lizard." Although some dinosaurs were indeed aggressive, fearsome, flesh-eating tyrants, many more dinosaurs ate only plants; others were no bigger than a robin! The creatures that inspire our imagination and awe were the immense, thunderous beasts that dominated the Earth's landscapes for some 160 million years. First popularized by dramatic museum displays of their bony skeletons, details of the life and habits of dinosaurs have been revealed to an eager public in cartoons, movies and books. Many youngsters can rattle off the tongue-twisting scientific names of any number of their favorite animals. For some people, the fascination with dinosaurs comes from trying to explain the mysterious disappearance of the giant creatures. Most *paleontologists*, however, are now convinced that dinosaurs can be seen daily at the backyard feeder: modern birds are the direct descendants of dinosaurs.

Two hundred million years ago during the Jurassic Period, dinosaurs were abundant in eastern North America. At that time, New England was located closer to the equator, and huge lakes existed in what is now central Connecticut and Massachusetts. Plants thrived along the lakeshores, and the waters supported large populations of *fishes* and *invertebrates*.

Dinosaurs and a variety of other animals were drawn to the lakeshores in search of food and water. Knowledge of the Jurassic life of the region comes from the fossil record preserved in the local rocks. These rocks were once sand, silt and mud in ancient lakes and streams.

The chapters that follow not only describe the world-class trackways preserved at Dinosaur State Park and the animals that probably made them, but they also provide details of the geology, climate, and varied plant and animal life of Jurassic times. Park exhibits include some of the best fossils unearthed from the local rocks. The murals, models, dioramas and displays in the Exhibit Center effectively re-create Jurassic environments, and provide clues as to how plants and animals interacted and survived in the days of the giant dinosaurs. The "Arboretum of Evolution" surrounding the Exhibit Center contains more than 200 plant types whose close ancestors thrived when large dinosaurs roamed the Earth. Exploration of the swamp, forest, meadow and rocky ledge *habitats* along the Park's nature trails invites comparisons of Jurassic and modern environments. Then, as now, organisms faced ongoing challenges to their survival. These challenges included competition for food, living space, and the ability to produce offspring, as well as the struggle to adapt to changes in landscape and climate. A visit to Dinosaur Park offers more than a glimpse into the distant past; it can motivate us to consider our personal and collective role in preserving life on this fragile planet.

Figure 1-2. One of the earliest views (looking west) of excavations to uncover track-bearing layers at the site that would become Dinosaur State Park.

CHAPTER 2

Time and Change

Human time is measured in minutes, hours, days and centuries, but these time intervals seem insignificant when compared with the immense age of the Earth.

Geologic time extends from the formation of the Earth right up to the present day. Geologic time is typically measured in millions and even billions of years. Modern geologists agree that the Earth is a very old planet, but accurate numerical estimates of the age of the Earth have been available for only about 50 years. These reliable estimates have come from studies of radioactivity. In simple terms, radioactivity can be defined as the loss of particles (protons, neutrons, electrons) and energy from unstable atoms (elements). Through a process known as "radioactive decay," radioactive elements shed particles at a very steady rate until they become stable elements. For example, some atoms of uranium, the best-known radioactive element, will lose particles steadily over time until they become atoms of a stable, non-radioactive element, lead. By measuring the relative amounts of radioactive and non-radioactive elements found in some rocks, scientists can estimate how long ago the rock was formed. Using radioactive dating methods, the age of the Earth has been determined to be about 4.6 billion years.

Figure 2-1. The first four billion years of Earth history is often called "Precambrian" time, and is represented on the time-line walkway by 80.6 feet of light gray granite. Each foot on the time-line equals 50 million years. Fossils are absent or sparse in Precambrian rocks; long, blank sections of the time-line reflect gaps in the knowledge of this earliest time.

The Geologic Time Scale and Time-line Walkway at Dinosaur Park

In the same way that "human" time is subdivided into centuries, decades, years, and so forth, geologic time has been subdivided into time intervals known as *eons, eras* and *periods*. Eons are the largest units of geologic time, sometimes lasting billions of years. Eons are divided into eras, the longest of which includes more than 300 million years; eras are broken down into periods. No two eons or eras cover the same length of time; some are longer than others, and the same is true for periods. This is so because the units of geologic time are separated from one another by important geologic events, or more commonly by changes in the fossil animals and plants found in rock layers. Divisions between eras and periods often reflect dramatic changes in the dominant organisms present in the rocks.

At Dinosaur Park, visitors can easily visualize the great expanse of geologic time by walking along the time-line that has been built into the sidewalk leading from the parking lot toward the main entrance of the Exhibit Center. The time-line is made of different colors of smoothed granite slabs embedded into the cement of the walk, and it displays the major eras of the geologic time scale. For simplicity, eon and period names are not shown, but the names of the various eras are provided on a descriptive bronze plaque at the start of the walk. The time-line is 92 feet long, and it represents the entire 4.6 billion-year history of the Earth. Each foot on the time-line thus equals 50 million years! The longest section of the time-line represents Precambrian time: the four billion years from the origin of the Earth to the appearance of abundant fossil life in the world's oceans (Figure 2-1). Large portions of this section of the time-line are blank, because our knowledge of this most ancient time is very incomplete. The oldest Precambrian rocks contain no trace of fossils. Younger Precambrian rocks do contain evidence of fossil life, beginning with single-celled organisms, such as bacteria,

and stromatolites, mound-like structures made by marine bacteria or algae (Figures 2-2 and 2-3). However, most animals living during later Precambrian time were simple and soft-bodied (such as jellyfish and worm-like creatures), and they were rarely preserved as fossils.

Figure 2-2. The oldest fossils on Earth are stromatolites, laminated sedimentary structures formed by colonies of marine algae or bacteria. These cross-sectioned, cauliflower-shaped stromatolites are exposed in early Paleozoic rocks near Saratoga Springs, New York. These were the first stromatolites described from North America. Leaves in the foreground provide scale.

Figure 2-3. Living stromatolite mounds can be seen today in the shallow, salty waters of Shark Bay, Western Australia. The larger mounds are about three feet high.

An "explosion" of new life forms initiated the Paleozoic Era, represented by 6.5 feet of pink granite on the time-line (Figure 2-4). The seas became populated with an amazing variety of life, including familiar animals such as mollusks, corals and early fishes, along with creatures now long extinct, such as *trilobites* and many varieties of *brachiopods*. Many of the ocean-dwelling animals had durable body parts, and fossils are common in rocks of this era. Later in the Paleozoic, plants began to colonize the land, and animals were soon to follow. Among the animals that first appeared at this time were *amphibians*, similar to modern-day salamanders and newts. Near the end of the Paleozoic, vast swamps developed, and huge forests of cone-bearing trees, gigantic seed ferns and *club mosses* covered the land. Coal deposits found in many parts of the world are the buried and preserved remains of these forests. A period of extensive desert conditions prevailed at the close of the Paleozoic Era, contributing to the greatest mass extinction in Earth history.

Reptiles became the dominant form of life on Earth during the Mesozoic Era, represented by about 3.6 feet of gray granite on the Park time-line. Reptiles ruled both land and sea, and one group, the dinosaurs, became the largest animals ever to live on land. Spreading *plates* led to the opening of the Atlantic Ocean early in the Mesozoic, and the sedimentary rocks and lava flows of central Connecticut were formed a short time later. The tracks at Dinosaur Park were made during the Jurassic Period, the middle of the three periods that make up the Mesozoic Era. Later in this era, flowering plants and birds made their first appearance in the fossil record, and *mammals* became more abundant. The Rocky Mountains were uplifted during late Mesozoic time. A dramatic extinction event brought the Mesozoic Era to a close about 65 million years ago.

Figure 2-4. The three most recent eras of geologic time, seen here at the far end of the walkway, comprise only one-eighth of Earth's history, but cover a time interval of more than 500 million years. Gray granite denotes the Mesozoic Era, informally called the "Age of Dinosaurs." The Park trackways date from the Jurassic Period, roughly 200 million years ago.

...ngest and shortest era of ...a in which mammals became ...mals. This era comprises only ...h granite on the time-line. ...*rimates* (lemurs and tarsiers) ...dest Cenozoic rocks; ...rimates) may have

originated as recently as five million years ago. Discoveries of human skeletal parts suggest that the first modern humans developed about 80,000 years ago. The time that modern humans have been on our planet can be measured by the thickness of a fingernail at the very end of the Park time-line!

Figure 2-5. A well-preserved dinosaur footprint from the Exhibit Center trackways at Dinosaur Park. These imprints are called "trace fossils" because they record evidence of an animal's activity, but do not preserve any body parts of the animal that made them. This track is 16 inches long, and one of the largest footmarks on display.

Fossils and the Record of Life on Earth

Fossils are evidence of past life, usually preserved in rock layers or in sediments such as sand, silt or mud. Many geologists consider that fossils must be at least 10,000 years old. Fossils are thus different from *remains*, which are evidence of past life less than 10,000 years old, and *artifacts*, which are any objects made or used by humans, such as arrowheads, carvings or clay pottery. Fossils can be the actual and original body parts of any animal or plant, but most fossils are physically or chemically altered (petrified) as a result of being buried underground. A special category of fossils called *trace fossils* includes tracks, trails, burrows and bore holes made by organisms, but such fossils show no actual body parts of the organism. The trackways at Dinosaur Park are trace fossils because they are the imprints of the feet of dinosaurs (Figure 2-5).

How does an organism become fossilized? The process starts with a living or recently dead plant or animal, or, at least, some of its parts. The organism then must be protected from the processes of decay and destruction. This usually means that the fossil-to-be is rapidly buried in an environment that will prevent it from being eaten or decomposed by other organisms, especially bacteria, fungus, insects and worms, and in an environment that will also protect it from weathering and erosion processes. Excellent environments for preservation include *anoxic* black mud that accumulates on the bottom of some lakes or in swamps, and the fine, limy sediments of ocean lagoons and shallow seas. Durable body parts such as shells, teeth, bones or woody tissue increase the likelihood of fossilization. However, even the most delicate of organisms, such as jellyfish and insects, can be preserved if the environmental conditions are right. Very few humans will ever become fossils, because our bodies normally are not preserved in suitable environments. It is no easy thing to become a fossil.

Figure 2-6. Cast of the earliest and most primitive fossil bird, *Archaeopteryx*, from Late Jurassic rocks in Germany. The size of a large chicken, *Archaeopteryx* has long been regarded as a "link" between reptiles and birds, but it has a greater resemblance to its dinosaur ancestors than to modern birds. Unlike today's birds, *Archaeopteryx* has a full set of teeth, a long, bony tail, three claws on each wing, and other features. Similar to modern birds, it has wings with flight feathers and a furculum (wishbone). Recent discoveries of feathered dinosaurs in China have solidified the evidence for evolutionary connections between some dinosaurs and birds. Most paleontologists now believe that birds are dinosaurs that became specialized for flight.

The fossil record provides abundant evidence that plants and animals have changed over time (Figure 2-6). All organisms on Earth face a struggle to survive; among organisms, there is always competition for space, food, and for the ability to reproduce. Plants and animals with traits that make them better able to compete will produce more offspring, and they will pass on those successful characteristics to their offspring. Accumulations of successful traits over generations may cause organisms to become quite different from their ancestors, and new *species* thus evolve. Organisms themselves have little influence on the direction of evolution. The concept of "natural selection" states that it is "nature" (the organism's environment) that determines whether an organism is fit or unfit. The fossil record provides many clear and compelling examples of evolution.

The fossil record also reveals that extinctions have played a major role in the history of life on Earth. The fossil record documents the complete disappearance of countless species of plants and animals, as well as entire groups of organisms, such as trilobites and many types of dinosaurs. Extinctions are brought about by a number of environmental conditions, including competition among and between organisms, disease, changes in climate, or catastrophic events (Figure 2-7). Whatever the cause, species that become extinct are the losers in the game of survival of the fittest. Extinctions influence the course of evolution because when species or groups of organisms die out, the environmental *niches* they once occupied are now available to other species.

Figure 2-7. Extinctions are well documented in the fossil record. Although it is still a much-debated subject, growing evidence suggests that the last of the non-avian dinosaurs died out after a large asteroid fragment (or fragments) struck the Earth about 65 million years ago. The discovery of a large, buried impact crater from that time on the Yucatán Peninsula of Mexico supports this theory. The impact would eject much rock debris and dust into the atmosphere, blocking sunlight and killing plant life. Large animals with large dietary needs, such as dinosaurs, would perish; smaller animals, especially those that could hide or hibernate, might survive this global catastrophe.

CHAPTER 3

Birth of a Valley

Dinosaur Park is situated within a long, broad valley.

This lowland extends through central Connecticut and Massachusetts from Long Island Sound to near the southern border of Vermont, a distance of more than 100 miles. It is about 20 miles across at its widest point. The Connecticut River now travels through much of this valley (Figure 3-1), but the lowland was created early in the Mesozoic Era, some 220 million years ago, long before the River began to flow.

At the beginning of the *Triassic Period*, the oldest of the three periods of the Mesozoic, all of the continents on Earth were joined into one immense landmass: a "supercontinent" known as *Pangaea* (Figure 3-2). North America was connected to Africa and Europe, and the Atlantic Ocean did not exist. Pangaea was created piece by piece, through the collision and fusion of plates during the Paleozoic Era. According to the concept of *plate tectonics*, the surface of the Earth is composed of at least eight major plates and several smaller ones. Plates are mobile, and their movement rearranges and reshapes continents and oceans over time. Driven by slow-moving currents of mantle rock deep below the Earth's surface, the average speed of plate motion is just a few inches per year, or roughly the speed at which your fingernails grow. The plate collisions that created Pangaea produced broad *mountain belts*, including the Appalachians of eastern North America.

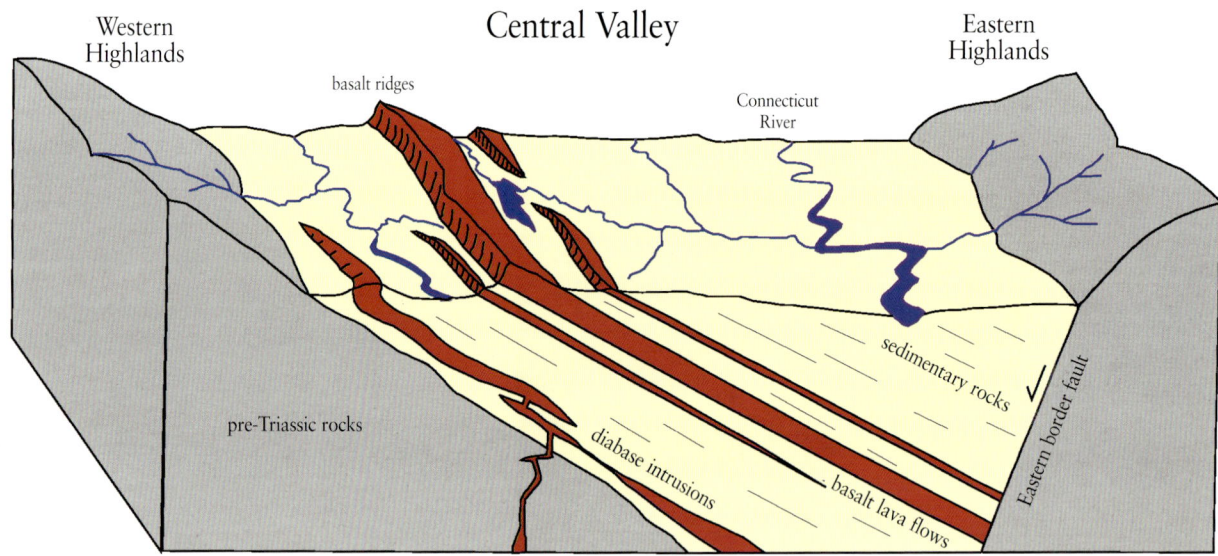

Figure 3-1. Simplified cross-section of the Central Valley of Connecticut. As one travels from west to east across the Valley, the Mesozoic sedimentary and igneous rocks exposed at the surface become progressively younger in age. *Basalt* and *diabase* are igneous rocks formed from lava and magma, respectively.

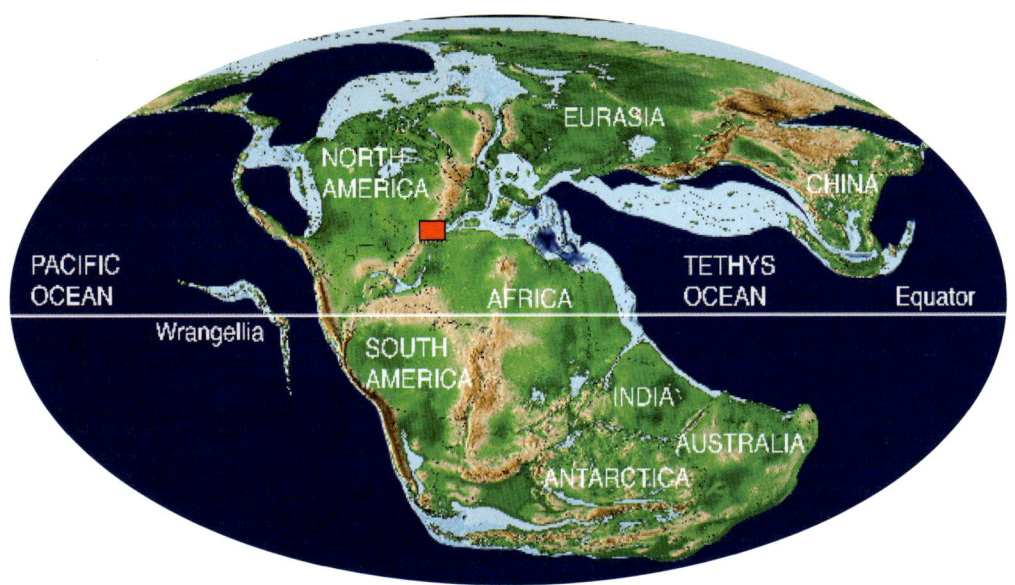

Figure 3-2. Idealized reconstruction of the "supercontinent" Pangaea early in the Mesozoic Era. Eastern North America and northwestern Africa were connected. The red box shows the location of New England.

Figure 3-3. A portion of Pangaea as it might have looked during Late Triassic times. Plate tectonic forces are beginning to separate Africa from North America, and the newly formed Atlantic Ocean is growing as the continents are rifted apart. The modern continental coastlines are outlined in black; Connecticut is highlighted in red.

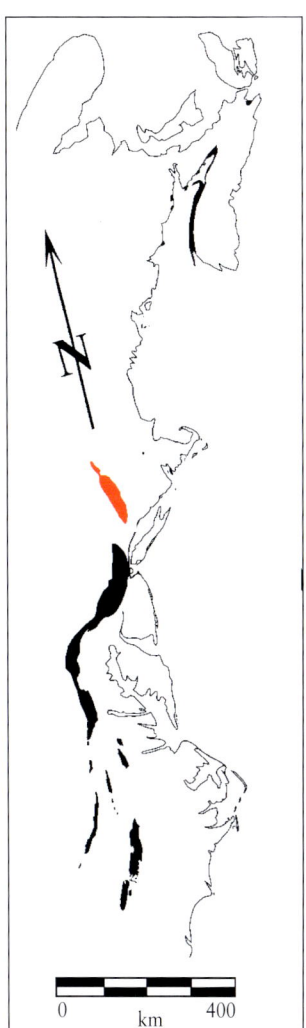

Figure 3-4. Map of some of the exposed early Mesozoic rift basins in eastern North America. The rift basins extend from Nova Scotia to South Carolina. The Central Valley of Connecticut and Massachusetts is highlighted in red.

Pangaea was in existence for about 100 million years, but late in the Triassic, plate tectonic forces began to slowly break the giant landmass apart. The breakup of Pangaea was caused by hot, rising currents of mantle rock, that first thinned, and then fractured and split the continental crust. Along the length of what would become our east coast, these powerful currents began to separate North America from Africa, creating the Atlantic Ocean through a process known as *rifting* (Figure 3-3). In the Late Triassic, the Atlantic was a long and narrow ocean, much like the Red Sea is today. Over time, it slowly grew larger, widening at a rate of about two inches a year. The Atlantic is still growing as plates continue to rift apart.

As North America and Africa were being stretched, torn and pushed apart, many zones of weakness developed. One of these zones of weakness ran along the east coast of North America from Nova Scotia to South Carolina. As stresses intensified, the crust along parts of the east coast broke along many long, deep *faults*, allowing blocks of the crust to sink, and forming a series of "rift valley" basins (Figure 3-4). One of these long, narrow depressions became the *Central Valley* of Connecticut and Massachusetts. Similar rift valleys are forming today in East Africa, where a part of that continent is gradually breaking away from the rest of Africa.

Geologists have estimated that the Central Valley continued to sink for as long as 35 million years, well into the Jurassic Period, due to continued fracturing on active north-south faults. As the rift basin deepened, rivers and streams eroded the low mountains and highlands to the east and west, filling the Valley with layer after layer of sediment: clays, sands and gravel. Like an immense, growing "layer cake," more than 20,000 feet of sediment accumulated in the Central Valley during the Triassic and Jurassic Periods. Over time, the sediments were buried, compressed and hardened into the rocks commonly seen in the region today: beds of reddish and black *shale*, and ledges of gray and brown sandstone and *conglomerate* (Figure 3-5).

In the earliest Jurassic, about 200 million years ago, an increase in the rate of rifting tilted the Valley floor gently to the east, creating long, narrow and deep lakes adjacent to the eastern highlands. These lakes existed for millions of years and are of great interest to paleontologists because of the abundant fossil life preserved in and around them. The oxygen-depleted black mud accumulating at the bottom of the lakes was an excellent environment for the preservation of fishes; plant remains that were carried or washed into the lakes also became fossils. Dinosaurs and other reptiles walked on the sandy shores of these lakes, and their footprints and trackways were often quickly buried and preserved. The spectacular trackways at Dinosaur Park were formed on the shoreline of a Jurassic lake, the gray beach sand having hardened into sandstone.

Some of the Early Jurassic faults in the region were deep enough to tap sources of magma in the lower crust or mantle, and several times molten lava flowed out onto the Valley floor. In places, the lava flows reached thicknesses of 450 feet! Most of the lava flows came out of *fissures*, great cracks in the crust, some as long as the Valley itself, or longer. Fissure eruptions are common in areas of crustal rifting, such as in Iceland today (Figure 3-6). Nearly all of the lava quietly oozed out or flooded out of the fissures in massive molten streams. The lava quickly cooled, forming a black, hard and dense igneous rock called *basalt*, also known as traprock. Basalt is a very tough and durable rock, and resists the forces of weathering very well, especially when compared with relatively soft sedimentary rocks such as shale and sandstone.

Figure 3-5. East-dipping beds of Early Jurassic red-brown sandstone and conglomerate exposed in roadcuts along I-84 near Manchester, Connecticut. The sediments, including rounded cobbles and boulders more than a foot in diameter, were transported into the Valley by powerful streams flowing from nearby highlands.

Figure 3-6. Clouds of volcanic gases and streams of basaltic lava flow out of actively spreading fissures during a 1977 eruption in the Krafla region of northeastern Iceland. Similar massive outpourings of lava took place in the Central Valley during the Early Jurassic.

In the Early Jurassic, there were three major intervals when fissures were active and when lava covered large parts of the basin and surrounding areas. When the eruptions ceased, more than 15,000 feet of mud, silt, sand and gravel from steams and lakes buried the hardened lava flows.

Later faulting tilted the layers of sediments and igneous rocks even more to the east, giving them an average easterly *dip* of 10 to 15 degrees. Activity on faults cutting across the basin broke and split the Valley deposits and lava flows into distinct fault blocks, exposing the buried layers. The cross-sectioned rock slab in Figure 3-7 provides an excellent visual example of this "block faulting." Compare this slab to the cross-section diagram of the Central Valley (Figure 3-10).

Figure 3-7. This sectioned slab of black shale and siltstone from Durham, Connecticut, shows deformation due to faulting. The originally horizontal rock layers have been offset by several "microfaults" and segregated into "blocks" bounded by faults. On a much larger scale, the Central Valley experienced similar structural deformation in the Early Jurassic. This slab measures 8.5 by 5 inches.

Figure 3-8. One of the best exposures of Jurassic lake cycles is revealed in Berlin, in roadcuts at the intersections of Connecticut Routes 9, 15 and 72. At these cuts, six cycles of lake expansion and contraction are readily identifiable by six evenly spaced units of black shale, each separated by about 40 feet of gray and red shale. Each cycle lasted roughly 20,000 years. Pictured above is part of the third cycle from the top at the Berlin roadcuts; these rocks may be the equivalents of those found at Dinosaur Park.

Climates of the Early Mesozoic

In the early Mesozoic, the immense landmass of Pangaea straddled the equator and was surrounded by oceans (Figure 3-2), and world climates were quite different from those of today. Levels of CO_2 in the atmosphere may have been four times higher in the Late Triassic and Early Jurassic than at present. This created a lengthy interval of elevated global temperatures, during which there was no ice in the polar regions and world climate zones were less extreme. Much of Pangaea probably experienced the effects of a strongly monsoonal climate, similar to that seen in the savanna grasslands of central Africa today. This is a climate with only two dominant seasons, a rainy season and a dry season, both of which last for months at a time.

Climate evidence from rocks and fossils also suggests that Pangaea slowly drifted to the north during the early Mesozoic. In the Late Triassic and earliest Jurassic, Connecticut was located in tropical regions some 15 to 25 degrees north of the equator. The oldest rocks in the Central Valley are mainly red sandstones and mudstones laid down by rivers that flowed from the highlands bordering the basin margins. In the Late Triassic, the climate was semi-arid, with a brief rainy season and an extended dry season. In the Jurassic, the deepening of the Valley and longer rainy seasons allowed for the development of lakes on the basin floor. These lakes varied greatly in size, depth and duration, from shallow, temporary ponds or *playas* to extensive, long-lived water bodies with depths of perhaps several hundred feet. The black, gray and red fine-grained sediments deposited in the larger lakes clearly indicate that these lakes not only expanded and contracted in response to seasonal changes, but also that lake levels fluctuated predictably over much longer time periods.

Regularly occurring variations in the Earth's orbit around the Sun affect the distribution of solar energy reaching the Earth. These orbital cycles greatly affect climate and can even initiate ice ages. Orbital cycles affected precipitation and thus lake levels in the Jurassic. In the rock record of the Central Valley, these orbital cycles produce cyclic layers of black shale (deeper lake deposits), gray shale (shallow lake deposits) and red shale (playa or stream deposits) (Figure 3-8). The different colors in the rocks reflect varying amounts of oxygen in the sediments. The rocks exposed at Dinosaur Park are shallow lake deposits associated with one of these cycles.

The youngest rocks in the Central Valley are red, brown and maroon beds of sandstone, mudstone and conglomerate deposited in river channels, on floodplains and on *alluvial fans* in the Early Jurassic. Extensive exposures of these *strata* occur in roadcuts along I-84 in the Buckland area of Manchester, Connecticut (Figure 3-5). Lake deposits are conspicuously absent

in these rocks, and this may indicate an increasingly dry climate as Pangaea drifted northward, entering more arid climatic zones. Molds and casts of salt crystals and other evaporite minerals are common in some rock units in the Valley. Sandstone beds formed from wind-blown sediment suggest that migrating sand dunes covered parts of the Valley floor at times. Mudcracked surfaces (Figure 6-12) occur abundantly throughout the early Mesozoic deposits of the Valley, recording short and frequent drying episodes on floodplains and lake shorelines. Some fine-grained rocks even contain the preserved impressions of raindrops from passing showers (Figure 3-9).

Figure 3-9. Raindrop impressions on fine red shale from Plainville, Connecticut. The largest drops are 0.25 inches across. These sedimentary structures are convex, thus they are mud-filled casts from the layer above the actual raindrop-pitted surface.

The Central Valley of Connecticut in Modern Times: The Rock Record

Major fault activity and deposition in the Central Valley ended by mid-Jurassic times. Since then, the region has experienced a long interval of gradual uplift along with massive amounts of erosion. The hardened, tilted and block-faulted "layer cake" of sediments and lava flows from the dinosaur age now can be seen at the surface, making up the ridges and valleys of the local landscape. This layer cake has been sliced by faults and tipped partly on edge, revealing large portions of the Triassic and Jurassic bedrock (Figure 3-1). The upturned edges of the basalt lava flows, which are very resistant to the forces of weathering and erosion, stand out as long, generally north-south ridges. The softer sandstones and shales that occur below, in between, and above the lava flows, are easily worn down by erosion, and form low hills and valleys. Although these sedimentary rocks are often not visible at the surface, they can be exposed in stream channels, along road cuts, or in quarries or excavations, as is the case with the rocks at Dinosaur Park.

The Triassic and Jurassic rocks in Connecticut and adjacent southern Massachusetts have been divided into seven separate *formations*, each named for a locality or region in which the formation is very well exposed (Figures 3-10 and 3-11). The oldest Mesozoic rock formation in Connecticut is the New Haven Formation. Except for its uppermost beds, the New Haven Formation is Triassic in age, and is composed mainly of reddish sandstone, siltstone and conglomerate deposited in river and stream channels and on river floodplains. Only a few fossils have been found within this formation, mostly skeletal parts of reptiles; no tracks or fish fossils are known from these rocks. This lack of fossils does not suggest that life was scarce during Triassic times in Connecticut, but rather that conditions were not ideal for the preservation of many fossils.

The Talcott Formation (or Talcott Basalt) overlies the New Haven Formation. It is the oldest of the three volcanic lava-flow formations in the Valley and is Early Jurassic in age, as are all the formations that lie above it. The fine-grained lake and stream deposits of the Shuttle Meadow Formation rest upon the Talcott Basalt. Well-preserved fossil fishes and plants are found in black shales (lake-bottom deposits) of this formation, and lake shoreline strata contain dinosaur tracks. The wall cases in the Park's

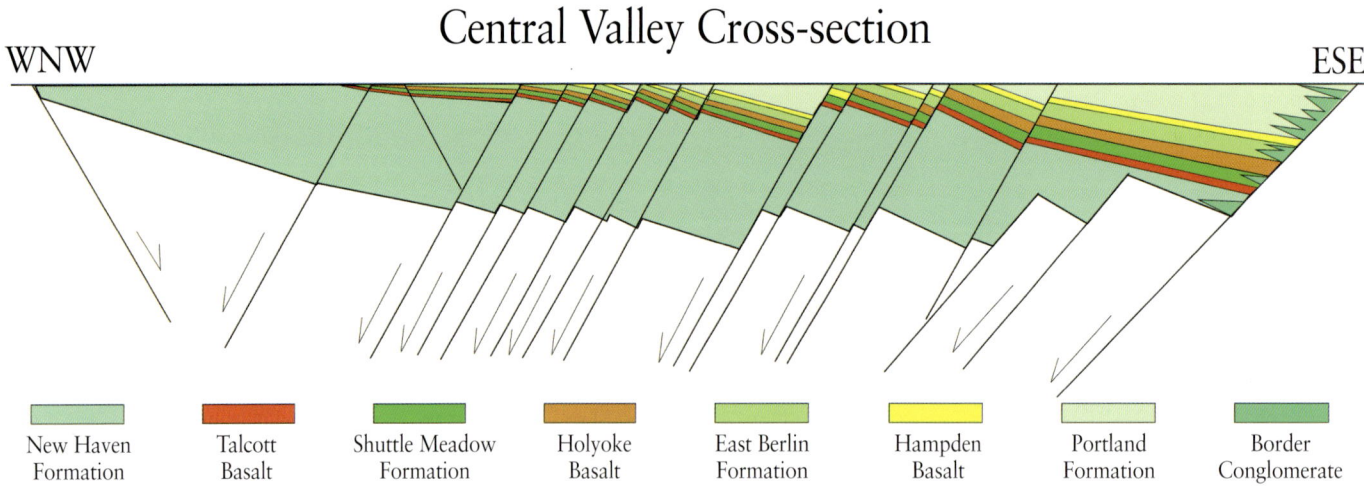

Central Valley Cross-section

WNW   ESE

| New Haven Formation | Talcott Basalt | Shuttle Meadow Formation | Holyoke Basalt | East Berlin Formation | Hampden Basalt | Portland Formation | Border Conglomerate |

Figure 3-10. Idealized cross-section of the Central Valley in central Connecticut, showing the early Mesozoic formations (colored) and underlying pre-Triassic "basement" rocks. Most apparent in this diagram is the subsidence, tilting and "block faulting" due to rifting. Arrows show the direction of movement along faults.

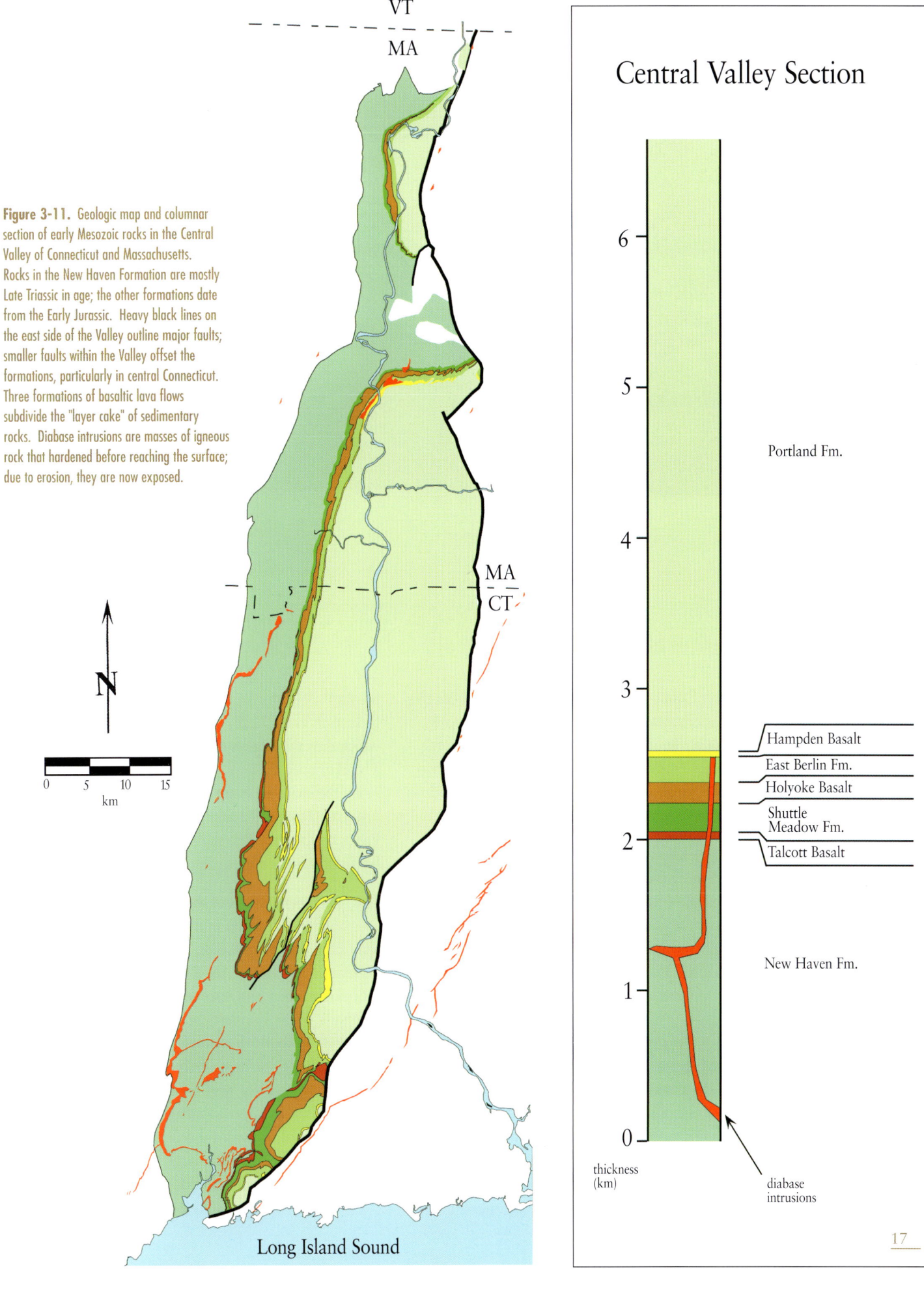

Figure 3-11. Geologic map and columnar section of early Mesozoic rocks in the Central Valley of Connecticut and Massachusetts. Rocks in the New Haven Formation are mostly Late Triassic in age; the other formations date from the Early Jurassic. Heavy black lines on the east side of the Valley outline major faults; smaller faults within the Valley offset the formations, particularly in central Connecticut. Three formations of basaltic lava flows subdivide the "layer cake" of sedimentary rocks. Diabase intrusions are masses of igneous rock that hardened before reaching the surface; due to erosion, they are now exposed.

VT

MA

N

0   5   10   15
km

MA
CT

Long Island Sound

Central Valley Section

6

5

4

3

2

1

0

Portland Fm.

Hampden Basalt
East Berlin Fm.
Holyoke Basalt
Shuttle
Meadow Fm.
Talcott Basalt

New Haven Fm.

thickness
(km)

diabase
intrusions

17

Figure 3-12. This two-inch-thick layer of lithified gravel from the East Berlin Formation, Branford, Connecticut, is sandwiched between beds of brown sandstone. The angular shape of the large quartz and feldspar fragments indicates that they have not traveled far from their source rock. Conglomerate with unworn, sharp-edged rock fragments is called "breccia."

Exhibit Center display a number of fossils from this formation, including a tooth from a carnivorous dinosaur. The Holyoke Basalt overlies the Shuttle Meadow Formation and is the thickest of the three basalt formations. It represents thousands of years of volcanic activity.

The East Berlin Formation forms a narrow belt of sedimentary rocks above the Holyoke Basalt in the central and east-central part of the Valley. The rocks of the East Berlin Formation (Figure 3-12) are very much like those of the Shuttle Meadow Formation, and also contain cyclic lake deposits. Most of these lake cycles are very uniform and continuous over a wide area, and some cycles contain abundant fossils. The sedimentary rocks and the spectacular trackways exposed at Dinosaur Park are within the East Berlin Formation, and are part of one of the conspicuous lake cycles in the upper portion of that formation (Figure 3-8). Other fossils from this formation, including plants, fishes and tracks, are on display in the Exhibit Center at

the Park. The third and youngest volcanic formation in the Valley is the Hampden Basalt, which directly overlies the East Berlin Formation. The Hampden Basalt forms a low, east-west ridge along the southern boundary of Dinosaur Park property (Figures 8-2 and 8-10). Several of the Park's nature trails travel along and through the basalt of this formation.

The sedimentary rocks of the Portland Formation, the youngest formation in the Central Valley, lie on top of the Hampden Basalt. The Portland Formation was named for the famous sandstone (popularly called *brownstone*) quarries along the east bank of the Connecticut River, in what is now Portland. In the 19th century, Portland brownstone became a highly fashionable building stone, and at one time the quarries were the largest in the world. Among the most impressive Park exhibits is a massive slab of brownstone from the Portland quarries containing several large *Otozoum* tracks, and showing the impression of the scaly skin on the dinosaur's feet (Figure 6-16). Fossils are abundant in some of the the lake-cycle units in the lower Portland Formation (Figure 3-13), and a few excellent specimens are on display in the Park's Exhibit Center. Quarry and roadcut exposures in the Buckland area of Manchester, Connecticut, reveal upper Portland rocks, and some of these strata have yielded the skeletal remains of dinosaurs. Other fossils from the Portland Formation at the Park are fossil clams from Suffield, Connecticut, and casts of dinosaur bones from East Windsor, Connecticut.

Although the Mesozoic rock record in the Central Valley ends with the Portland Formation, weathering, erosion and deposition have continued in Connecticut for the last 190 million years. Our landscape no longer experiences sputtering volcanoes or basaltic lava pouring out of long fissures, great tropical lakes expanding and contracting with the pulses of a monsoonal climate and teeming with primitive fishes, or the footfalls of the dinosaurs. Nevertheless, nature has left its legacy on the land.

Figure 3-13. Exposures of the Early Jurassic Portland Formation along Stony Brook, Suffield, Connecticut. The lower red rocks are stream-deposited sandstones and shales; the upper beds are gray sandstones and shales that formed in shallow lake waters. These lake beds have produced a variety of fossils including dinosaur tracks, fish scales and bones, plant remains, clams, and other invertebrates.

CHAPTER 4

Discovery!
Dinosaur Tracks Revealed

Tuesday, August 23, 1966, was no ordinary workday on his bulldozer for Edward McCarthy.

At a site on State of Connecticut property just south of West Street in Rocky Hill, McCarthy and his fellow construction workers were clearing ground and excavating for the foundation of a new State Highway Department Research and Testing Laboratory. While digging some 12 feet below the ground surface, McCarthy's bulldozer pulled up a large slab of gray sandstone. The sharp-eyed and curious operator noticed large, peculiar traces on the rock that looked like three-toed footprints (Figure 4-1). McCarthy pointed out the tracks to Thomas Jeffreys, an engineer for the Public Works Department, and to Thomas Perry, the architect for the new laboratory, who happened to be on-site that day. They immediately notified the University of Connecticut, the Peabody Museum at Yale University, and several local newspapers. Within hours, news of the discovery was reported on radio and television.

Digging continued the next day, and more track-bearing slabs were uncovered and pushed aside. On Wednesday evening, a large number of fossil enthusiasts and souvenir hunters armed with crowbars, hammers and chisels flocked to the excavation to collect specimens. One man is said to have started up the contractor's bulldozer so as to rip up more slabs. Among the visitors was Jane Cheney, Director of the Children's Museum in West Hartford, who

Figure 4-1. Bulldozer operator Edward McCarthy kneels beside an overturned gray sandstone slab containing a distinct three-toed footprint. Jurassic-age dinosaur tracks were discovered during excavations for a State building.

explored the site and examined more than 30 tracks and associated ripple marks and mud cracks. Alarmed by the potential destruction and loss of scientifically important specimens, Cheney contacted State officials, urging them to rope off the area and have it patrolled.

On Thursday, August 25, Public Works Commissioner Timothy Murphy Jr. asked Joe Webb Peoples, Director of the Connecticut Geological and Natural History Survey and Professor of Geology at Wesleyan University, to evaluate the scientific importance of the site and its fossils. Peoples met with Public Works and Highway Department officials on location that afternoon, accompanied by geologists and paleontologists from local institutions: Larry Frankel and Hugo Thomas from UConn, Henry Roos of Willimantic State College (now Eastern Connecticut State University), and John Ostrom and Elwyn Simons from Yale. The scientists were at once impressed with the scientific and educational potential of the locality. It was likely that well-preserved dinosaur footprints were present across much of the 14,400-square-foot area of the foundation excavation, and that durable, track-bearing sandstone layers continued downslope and to the west, still covered and protected by the overlying rock strata. The experts agreed that an exhibit of dinosaur trackways left in place would have great appeal to the public and would create a unique outdoor classroom. On learning of the unanimous opinion of the group of scientists, Commissioner Murphy issued an order that nothing more be removed from the site and that all earthmoving activity cease. Excavation work for the $1,067,000 Highway Department building came to an abrupt halt. By Friday night, a snow fence was installed around the excavation, straw and sand covered the exposed bedrock and loose slabs, and the site was patrolled around the clock by state troopers, later to be replaced by a private security agency.

The next few days saw a flurry of activity and debate about the future of the Rocky Hill track site. The main topic of discussion among State representatives and the geologists was whether to continue the construction project and relocate the best slabs and tracks to another location, such as a museum, or to convert the site into an open-air nature preserve where the trackways could be seen by schoolchildren and the general public. On Sunday, August 28, an editorial in the *Hartford Courant* proclaimed, "It will be something pretty wonderful if the dinosaur site can be preserved," and expressed hope for a new State park. The newspaper carried front-page reports about the discovery and the dinosaur trackways for 12 consecutive days. Cheney went directly to Governor John Dempsey (about to run for re-election) and convincingly asserted that the track site was unique and worthy of preservation. On August 29, during another meeting with Peoples, Ostrom, Frankel, Roos and others, State officials agreed to look for other possible locations for their Highway Department building; the Rocky Hill site would become a park. Less than a week had passed since Ed McCarthy's accidental but fortunate discovery.

Figure 4-2. With the aid of heavy equipment, construction workers, paleontologists and student volunteers carefully uncover the main track-bearing layers. The abundance of tracks at the site was just becoming apparent.

Figure 4-3. A worker examines a plaster cast of one of the better dinosaur footprint impressions on a recently exposed bedrock layer.

At the August 29 meeting, it was agreed to extend excavations at the site to the south and west, in order to expose more of the primary track-bearing layers. Grant Meyer, chief preparator at the Peabody Museum, was engaged to direct the work of uncovering and preserving the track slabs and bedrock trackways, assisted by staff from Yale and UConn and several graduate students. Peoples and Ostrom advised and supervised the project. Heavy equipment was used to raise and transport massive sandstone blocks (Figure 4-2). Hand tools, chisels and brushes were employed for the fine and delicate work of completely uncovering the specimens (Figure 4-3). Some tracks were treated with epoxy to resist fracturing. Clearing by teams of professionals and students continued until approximately 1,500 tracks were exposed on the surface of the main track horizon, and it became obvious that the track-bearing layers were much more extensive (Figure 4-4). In a September 7 letter to the State, Meyer wrote, "It is my opinion that this site should be preserved *in situ* for it is, I believe, without equal in the United States. Nowhere, to my knowledge, are so many dinosaur footprints found in such a small area. The value of this site lies in the fact that we have such an amazing density of footprints." On September 13, Governor Dempsey announced that the 7.7-acre site that contained the tracks would be preserved and designated "Dinosaur State Park." Effective communication, a rapid response, and a remarkable degree of cooperation between State officials, scientists and other interested parties rescued the tracks from immediate damage and preserved them as a public resource.

(right) **Figure 4-4.** As news cameras roll, Joe Webb Peoples describes the cleared portion of the trackway exposures to members of the press. The bedrock layers are tilted toward the south. West Street is at upper right.

(below) **Figure 4-5.** Perched high on a ladder mounted to the front of his Land Rover, Sidney Quarrier, a State geologist, painstakingly photographed most of the original trackway area, creating the photomosaic in Figure 4-6.

Figure 4-6. More than 1,500 tracks are visible in this 1969 photomosaic of the original trackway area, located east of the current Exhibit Center at Dinosaur State Park. A prominent set of fractures in the bedrock trends north-northwest. More than 80 distinct dinosaur trackways can be seen on this photomosaic; trackways lead in almost all directions. These trackways have been covered over for their protection since 1978. At lower right is a *Eubrontes* track with a quarter for scale.

(left) **Figure 4-7.** In June 1967, a new area of bedrock was uncovered west of the original track discovery site. Hundreds of additional footprints were revealed, and an inflated nylon fabric dome was built over the new trackways.

(below) **Figure 4-8.** The first Park building, nicknamed the "bubble building," was held up by air pressure. This temporary building was to be replaced by a much larger structure that would enclose the entire track area. The original trackway site is in the right foreground.

At the end of October 1966, the decision was made to suspend the excavation operations, as there were growing concerns about how to protect the exposed trackways from the ravages of New England's winter weather. Freezing temperatures and the resultant expansion of moisture in the large and small fractures that crisscross the trackways might cause extensive damage. Accordingly, three sump pits were dug and equipped with pumps to improve surface drainage and to lower groundwater levels. Students from Wesleyan, Yale and UConn helped to coat the surface of the main trackway layers with a plastic sealant, and the tracks were then covered by sheets of vinyl plastic, electric heating cables on a wooden grid frame, layers of sand and mulch, and finally a cover layer of vinyl, all weighted down by used automobile tires.

The following April, a landscape architectural firm presented an elaborate master plan report to the State Park and Forest Commission, outlining design and development ideas for the Rocky Hill site. The company's recommendations included construction of a large, rectangular exhibit building over the trackways, complete with balconies surrounding the interior of the structure, allowing both space for exhibits and for viewing the footprint surface from several aerial vantage points.

Other plans included the purchase of additional land, the building of supplementary exhibit halls and a curator's residence, and a large parking lot north of West Street connected by a pedestrian overpass to the track site. Also recommended was a series of nature trails, to enhance the ecological study of the woodland, marsh and pond areas located just south of the planned exhibit buildings. Unfortunately, the price tag for this grand plan topped more than six million dollars, and it did not meet with State approval.

Later in the spring of 1967, the State Legislature did authorize $250,000, part of which was to be used for the construction of a temporary exhibit building and support facilities at the Park. Since the large area of tracks covered by the "electric blanket" was to be preserved for a future, permanent building, excavations began in June to expose a new area of tracks just west of the buried trackways. Sidney Quarrier, a geologist with the Connecticut Geological and Natural History Survey, supervised the excavations that summer, and also was charged with developing the educational programs and exhibits for the Park. By the fall, a fresh exposure of bedrock (Figure 4-7) revealed an additional 600 well-preserved tracks, and construction of the building to house the new trackways began early in 1968. The new building, an inflated, plastic-coated, white nylon fabric dome, was completed late in the summer of that year. The inflatable balloon, which came to be known as the "bubble building," was hailed as an ultra-modern, "space-age" facility (Figure 4-8). The Park was officially opened to the public on October 17, 1968. At the opening ceremony, the National Park Service presented a bronze plaque recording that the Dinosaur Trackway had been designated as a Registered Natural Landmark. The Department of the Interior classifies Natural Landmarks as sites that possess exceptional value in illustrating the natural history of the United States.

The newly opened Dinosaur State Park, with its mid-state location and easy access to interstate routes, immediately became a very popular destination for State residents, tourists and school groups. Full-time staff and trained guides were employed to aid the public in interpreting and appreciating the Park's exhibits. In the summer of 1969, the buried trackway area was again uncovered (Figure 4-6), and the public reception was remarkable. In both 1969 and 1970, the number of visitors exceeded 100,000. In July 1970, Richard Krueger joined the staff and assumed responsibility for the instructional programs at the Park. Trained as a geologist, but with broad interests in botany, zoology and *ecology*, Krueger immediately began to utilize the tremendous educational potential of the Park's natural surroundings. The study of living plants and animals became a regular part of Park programs. Aided by Park staff, Krueger initiated the design and construction of nature trails to connect the trackway exhibit areas with the variety of natural habitats on the grounds to the south. A "Mesozoic garden," containing living representatives of primitive plants, was established near the exhibit building.

However, all was not well with the bubble building. On hot summer days, temperatures within the enclosure could reach 120 degrees. In the spring and fall, condensation on the inside of the dome produced drips of water. Guides were challenged by the acoustics of the building when describing the trackways and exhibits to groups of visitors.

A severe storm in December 1969 collapsed the bubble and tore a large hole in the fabric at the north end. Throughout the rest of the winter, the structure remained frozen to the ground. It was patched and reinflated in May 1970, but the skin of the building was abraded and badly discolored where it had been in contact with the ground. Over the next few years, the status of the bubble building as a "temporary" structure became quite literally true; collapses under the weight of winter snow were a routine occurrence. Staff and supporters of the Park hoped that resources for the construction of a permanent building to cover the entire trackway area might soon be forthcoming. The tracks outside the bubble building were covered by an easily removable "sandwich" of wood chips and plastic sheeting, held down by bales of hay. Following the final collapse of the bubble in 1975, the Park remained closed for almost three years, awaiting the design and construction of a new and more durable building. An insulating layer of straw and fabric salvaged from the bubble was laid down to protect the tracks formerly inside the building. Work continued on the nature trails and outdoor exhibits.

A new Exhibit Center was built on the site of the bubble building and opened to the public in 1978. The striking building is pentagonal in shape and

Figure 4-9. The Exhibit Center in the early 1980s. Capped by an aluminum geodesic dome, this structure houses approximately 600 dinosaur tracks in place on gray sandstone bedrock. The building is now surrounded by the attractive landscaping and plantings of the "Arboretum of Evolution."

25

capped by an impressive aluminum geodesic dome, 52 feet high at its center (Figure 4-9). As well as protecting and displaying some 600 *Eubrontes* tracks in place on their original bedrock layers, the Exhibit Center houses a 100-seat auditorium, a classroom/demonstration room, a bookshop, and work areas for staff and volunteers. The *Eubrontes* trackways in the Exhibit Center are surrounded on nearly all sides by dramatic murals, dioramas, life-size dinosaur models and by some of the best fossils ever found in the Jurassic rocks of the Central Valley. As the new Exhibit Center neared completion, the large area of trackways outside received a more durable covering: polyethylene sheeting, a sprayed application of two inches of polyurethane foam to conform to the trackway surfaces and provide uniform cushioning, four feet of sand and an upper layer of topsoil. More than 1,500 tracks remain buried and protected under the lawn immediately to the east of the Exhibit Center.

A gift of dawn redwoods to the Park in 1979 marked the humble beginning of the Arboretum which now occupies the ten acres surrounding the Exhibit Center. The Arboretum, largely the result of the vision and efforts of former Park director Krueger, is unique in the State for the variety of trees and other plants it contains. Seeking to create plantings that would resemble the ancient forests in Connecticut during the Early Jurassic, Krueger solicited donations of firs, spruces and other *conifers* from nurseries and other sources, grew bald cypresses and other trees from seedlings, and nurtured arborvitae from cuttings. Currently, the Arboretum contains more than 200 species and *cultivars* of conifers, as well as katsuras, ginkgoes, magnolias and other living representatives of plant families that first appeared during the Age of Dinosaurs. There are also attractive gardens of roadside flowers, native wildflowers and grasses, a rock garden, and a butterfly garden.

The nature trails began as a single woodland pathway and boardwalk in 1970. They now comprise more than 2.5 miles of marked walkways, made possible by donations and acquisition of land south of the Exhibit Center and by the diligent clearing and maintenance efforts of Park staff and volunteers. The trails link the Exhibit Center with

natural areas that contain a rich diversity of plants and animals in a variety of habitats. The habitats include a red maple and shrub swamp; forested areas of sugar maple, birch, hickory, oak and beech; open meadows; and a basalt (traprock) ridge complete with a boulder-strewn *talus slope*, and a natural spring. Informative signs highlight the evolutionary history of animals seen throughout the Park, and illustrate the geologic history of the area.

The Exhibit Center Trackways

Since their discovery, the world-class exposures of Early Jurassic dinosaur trackways at Rocky Hill have drawn eager visitors to the Park, from the youngest schoolchildren to renowned research scientists. Under the gleaming geodesic dome of the Exhibit Center, an 11,000-square-foot area of exposed bedrock contains some 600 tracks, preserved in gray sandstones that once were the beach sands of a lake shoreline 200 million years ago. Shadows cast by a combination of low-angle floodlights and spotlights from above highlight the large, deeply etched, three-toed tracks (Figure 4-10). Visitors first see the gently inclined trackways from a high vantage point overlooking the tracks near the entrance of the building. The eye naturally follows the stony trackways up-slope. In the background, the trackways blend with a diorama and mural of a Jurassic lakeshore, complete with a life-size model of the probable track maker, the carnivorous dinosaur *Dilophosaurus*. Proceeding counterclockwise, passing exhibits and displays describing the history of the Park and the geology and fossils of the Central Valley, visitors descend to a glass-sided walkway, elevated just a few feet above the trackways. Here, you are within a stride of a

(above) **Figure 4-10.** Part of a trackway inside the Exhibit Center, likely made by an Early Jurassic carnivorous dinosaur about 20 feet in length. The ripple marks on the bedrock surface are indicative of very shallow water.

(right) **Figure 4-11.** Six feet of sedimentary rock was intentionally left in place above the main trackway beds in one corner of the Exhibit Center. A well-preserved footprint is visible in the foreground; the rest of this dinosaur's trackway remains buried beneath the rock layers.

realistic *Dilophosaurus*, 20 feet long and six feet tall at the hip. Just as you come face-to-snout with the imposing flesh-eater, your movement activates an audio and video program that simulates the sounds of the forest and the stomping and huffing of the big dinosaur. The lights dim as a thunderstorm builds in the distance. Walking between the expansive trackways and the reconstructed ancient lakeshore and forest, you can imagine that you are in a Jurassic landscape on a humid summer day.

The most frequently asked question about the trackways is "Are they real?" The answer is "Yes!" Figure 4-7 shows the uncovering of the trackways now visible in the Exhibit Center during the summer of 1967. The tracks are imprinted on tilted layers of sandstone that extend underground for an unknown distance to the south, west and east. The sandstone beds are part of the East Berlin Formation, a sedimentary rock formation composed mostly of lake and stream deposits.

Like the larger area of now-covered trackways just east of the Exhibit Center, the tracks inside the building were found some 12 feet below the original ground surface. To enable Park guests to better visualize that the footprint-bearing layers were buried beneath numerous other beds of sedimentary rock, the designers of the Exhibit Center left a small portion of bedrock in place at the northwest corner of the trackway area. In this area, about six feet of sandstone, siltstone and mudstone can be seen above the trackway layers (Figure 4-11). The color, particle size and other features of the rock layers still exposed above the trackways reveal much about the Early Jurassic lake and stream environments in which the original sediments were deposited (Figure 4-12).

During the initial excavations at the Park site, dinosaur footprints were observed on five distinct layers within the ancient shoreline deposits. In nearly all of these track-bearing layers, however, the footprints were poorly preserved, the beds split unevenly, or the rock was crumbly or brittle. Fortunately, Ed McCarthy's bulldozer split apart a thick, hard and blocky unit of sandstone with abundant and well-preserved tracks. Approximately one-fourth of the originally exposed surface of this durable sandstone unit is now on display under the dome (Figure 4-16). In the Exhibit Center, dinosaur tracks can be seen on two beds of sandstone, one lying just above the other. The upper bed, containing most of the footprints, averages 1.5 inches in thickness and is composed of light gray, fine-grained to coarse-grained sandstone. The lower sandstone bed, visible near the walkway area, is about two inches thick, and is similar to the upper bed in color and composition. The lower bed can easily be identified by its ripple-marked upper surface. Both of these sandstone beds contain zones rich in silt, clay and mica where the rock splits easily.

All but two or three of the footprints in the Exhibit Center trackway area are the large, three-toed tracks known as *Eubrontes*. Fossil tracks are given their own scientific names because they are rarely found associated with bones. No dinosaur bones have been found at the Park. *Eubrontes* tracks likely were produced by a *theropod* dinosaur similar to *Dilophosaurus*, though it is unclear whether the *Eubrontes* tracks represent a single species or several closely related species. A smaller track, which has been given the name *Anchisauripus*, is present on the upper sandstone bed next to the walkway. *Anchisauripus* tracks may have been produced by younger individuals of the *Eubrontes* track maker, or by a smaller theropod relative.

All of the *in situ* tracks in the Exhibit Center are the actual imprints of dinosaur feet, or, more precisely, toes. These impressions are referred to as "positive" tracks. Large sandstone slabs bearing

C. Lake shallowing

B. Lake at maximum depth

A. Lake expansion and deepening

Figure 4-12. The six-foot thickness of bedrock preserved in the Exhibit Center records three phases in the history of an Early Jurassic lake. The lowermost beds, consisting of light gray to tan siltstone and sandstone, formed in shoreline regions while the lake was expanding and gradually deepening (phase A). Ripple marks, mud cracks and dinosaur tracks found in some of these beds indicate extremely shallow water or beach conditions. The layers of gray-black shaly mudstone and dark gray siltstone in the center of the exposure formed from fine-grained clay and silt particles deposited in offshore waters when the lake was at maximum depth (phase B). The darker color of these rocks reflects low oxygen levels in the deeper water and the preservation of organic matter in the sediments. Beds of light gray and tan sandstone near the top are composed of shoreline sediments laid down as the lake shallowed and contracted (phase C). At the very top of the exposure, mostly hidden from view, are beds of maroon sandy siltstone; these rocks formed from oxygen-rich stream sediments. The three phases may represent a time interval of up to 20,000 years.

raised "negative" *Eubrontes* tracks can be seen outside in the parking lot, along the path leading to the nature trails, and especially in the casting area. These slabs came from a thick, blocky layer immediately above the upper sandstone bed under the dome. Heavy equipment was used to pry up slabs of this sandstone. Due to the mica-rich nature of the rock, most positive imprints and negative "out-prints" cleanly separated from one another (Figure 4-13). However, the layers did not always split perfectly. In the trackway area, there are numerous roundish-to-oval patches that look rather like "pancakes" on the surface of the upper sandstone bed. These pancakes are places where positive and negative tracks did not part cleanly; the positive impressions are still filled with rock from the overlying layer. Some of the pancaked patches were carefully cleaned out with hand tools, chisels and brushes to reveal the concealed footprints.

Most of the *Eubrontes* footprints on the upper sandstone bed are quite well preserved and show the impressions of three toes very clearly, but some prints are deeper and more sharply outlined than others. Because the prints were made in fairly coarse sandy sediment, few tracks show claw marks or distinct impressions of the pads on the underside of the foot; skin impressions have not been observed on any of the *Eubrontes* tracks. Several explanations can account for the range in the quality of footprint preservation on the upper trackway bed, and experts have not come to any agreement as to how much time it took for so many tracks to accumulate on the same surface. The shallow track impressions may be *underprints*, but more likely they result from footmarks made hours, days or even weeks earlier than the crisp, well-formed impressions. Some of the faint tracks are overlapped by younger, deeper tracks, and this suggests the passage of time. The consistency (hardness or softness) of the sandy sediment also affected preservation. Tracks made in shallow water, or on a puddle-covered surface,

or just after a rainstorm, would look different from footprints made on a dry sand flat that had been baking in the sun (Figure 4-14).

Shallow, indistinct *Eubrontes* impressions are found on the lower sandstone bed, exposed only at the northern end of the trackway area (Figure 4-16). Most of these tracks are underprints. The broken, exposed edge of the upper sandstone bed cuts through several *Eubrontes* tracks, and impressions of the same footprint can be seen on both the upper and the lower sandstone bed. In several cases, successive footprints made by the same animal can be followed from the upper sandstone to the rippled surface of the lower sandstone; tracks that are deep and well outlined on the upper bed turn into faint underprint traces on the lower bed.

Figure 4-13. Positive and negative *Eubrontes* track slabs inside the Exhibit Center. On the right, a "positive" track is the actual indented mold the dinosaur's foot made in the sediment, now hardened into rock. At left, a "negative" track is a raised cast of the sediment which filled in the original impression. Due to the mica-rich nature of the rock, most positive and negative tracks readily separate from one another. The track is 15 inches long.

Figure 4-14. These *Eubrontes* tracks on a now-buried rock layer east of the Exhibit Center illustrate how sediment consistency affects print quality. At left, the dinosaur stepped on a firm sediment surface, creating shallow, indistinct toe impressions. As the animal walked into deeper water and softer sediment, the footprints became deeper and the toes more clearly defined. The ripple marks outline the water's edge. The *Eubrontes* tracks are about 11 inches long; smaller, faint *Anchisauripus* tracks proceed from lower right to upper left.

Figure 4-15. *Eubrontes* trackways on the original exposures. The tracks outlined with chalk are about a foot long and ten inches wide; successive footprints are typically 3.5 to 4.5 feet apart. Short step lengths indicate that these dinosaurs were walking rather than running.

The size and spacing of the tracks provide clues to the appearance and behavior of the animals that left them behind (Figure 4-15). The *Eubrontes* tracks in the Exhibit Center range from 10 to 16 inches in length and from 10 to 12 inches wide. Successive tracks are spaced 3.5 to 4.5 feet apart. These measurements suggest that the adult track makers stood about six feet tall at the hip. Short step lengths imply that most animals were walking rather than running. More than 30 separate *Eubrontes* trackways have been identified on the upper sandstone bed under the dome. Typically, the individual trackways are laid out with one print directly in front of another, in a nearly straight line. This indicates that the track maker walked with its hind limbs held close to its body, not extended out to the side like a lizard. No definite forefoot impressions have been recognized on the Exhibit Center trackways, nor are there any tail-drag marks associated with the *Eubrontes* tracks. The *Eubrontes* track maker was clearly *bipedal*, and must have carried its stout tail off the ground, to counterbalance the forward-leaning front of its body.

Several researchers have investigated the orientation of the numerous *Eubrontes* trackways at Rocky Hill, mainly to determine if the dinosaurs making the tracks traveled together as a group. Working from the photomosaic of the buried trackways east of the Exhibit Center (Figure 4-6), John Ostrom (Yale) measured the orientation of 86 trackways and a random sample of 50 isolated *Eubrontes* footprints. Ostrom's data reveal trackways heading in almost all directions, but he also recorded several "clusterings" of trackway orientations, suggesting a number of preferred directions of travel for the track makers. He concluded that the "random wanderings" preserved in the sandstone at Rocky Hill were made over an extended period of time.

He noted that, given sufficient time, the comings and goings of several "herds" of dinosaurs across the same area would mostly obscure the trackway orientations made by any single group. Thus, the evidence from the Park trackways did not rule out social activity among the dinosaurs. In contrast, at the Mount Tom site near Holyoke, Massachusetts, Ostrom observed that 19 of 22 preserved *Eubrontes* trackways were straight and led in the same direction. He concluded that the Mount Tom trackways were made over a short interval of time by a herd of dinosaurs with a seemingly purposeful travel destination.

A travel-direction analysis of some 30 trackways in the Exhibit Center by James Farlow (Indiana University) and Peter Galton (formerly at the University of Bridgeport) arrived at essentially the same conclusions as did Ostrom. The trackways under the dome also are oriented in all directions of the compass, with some clustering toward the southwest, west and northerly directions (Figure 4-17). Farlow and Galton argue that there is no obvious indication that the Park trackways were made by a large group of animals moving through the area together. Instead, they envision individual dinosaurs, or possibly small groups of dinosaurs, passing across the site over an extended time, perhaps weeks or months. Given that the climate of the region is thought to have been monsoonal during the Early Jurassic, preservation of tracks for months on an exposed and expanding lake shoreline during a long dry season is quite possible.

Among the most unusual and most studied features on the Exhibit Center trackway beds are small, lightly indented claw marks on the surface of the upper sandstone bed, clearly made by a moving reptile. These curious markings (some 50 in number) were first noticed by Richard Krueger and later described by Walter Coombs Jr. (Western New England College) in a 1980 journal article. Typical footmarks of this kind consist of impressions of the tips of three toes (Figure 4-18). Some of these three-toed traces are shallower and fainter than others, some are incomplete, and some have been overprinted by typical *Eubrontes* tracks. Many of the claw marks are single, isolated prints or they

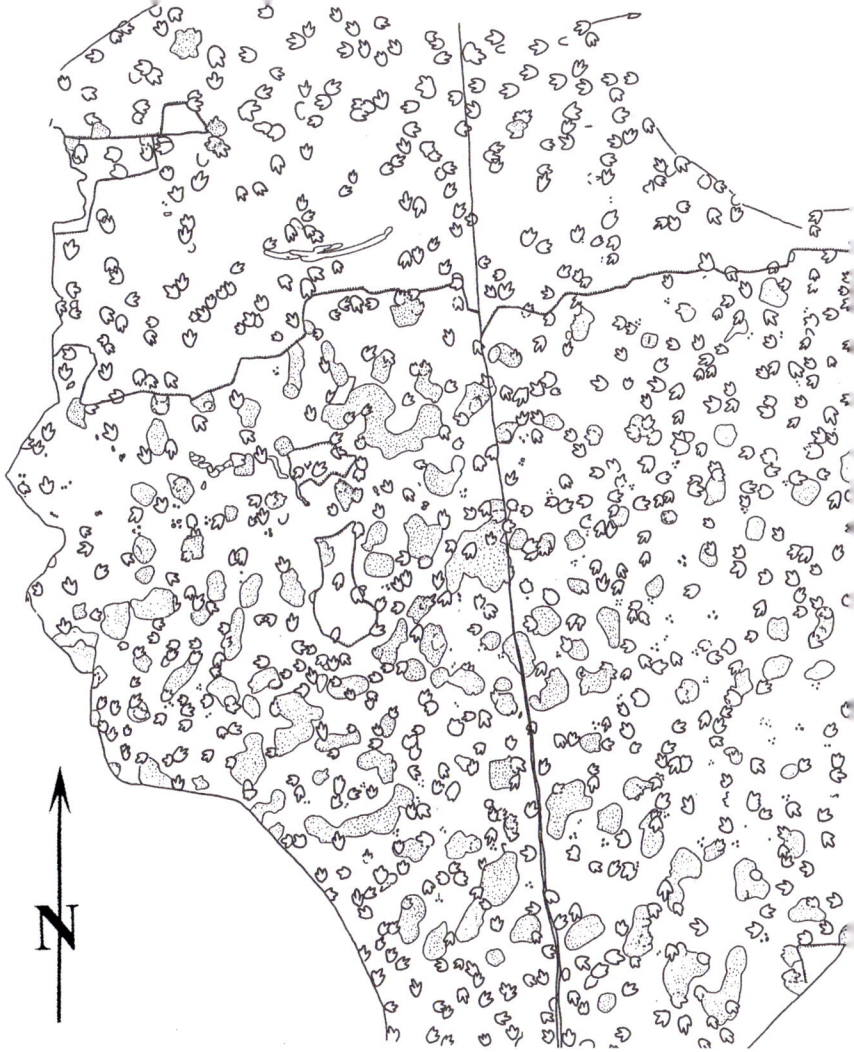

Figure 4-16. Sketch map of the estimated 600 *Eubrontes* footprints in the Exhibit Center. The track impressions range from 10 to 16 inches in length. Stippled areas indicate where positive and negative tracks did not part cleanly, leaving positive impressions filled with rock from the overlying layer.

start and end abruptly, but one sequence of markings consists of eight consecutive right and left footfalls made by the same animal. Coombs interpreted these markings as "swimming tracks," made by a partially submerged theropod pushing itself along in five to eight feet of water by digging the tips of its toes into the lake bottom (Figure 4-19). Based on the spacing of the toes, he suggested that the markings were produced by a theropod dinosaur similar in size to the *Anchisauripus* track maker, or by small or young individuals that made the *Eubrontes* tracks.

In their study of the trackways at Dinosaur State Park, Farlow and Galton also re-examined the nature and origin of Coombs's "swimming tracks." While acknowledging the possibility of swimming theropods, these authors suggested that the unusual markings were more likely shallow claw imprints

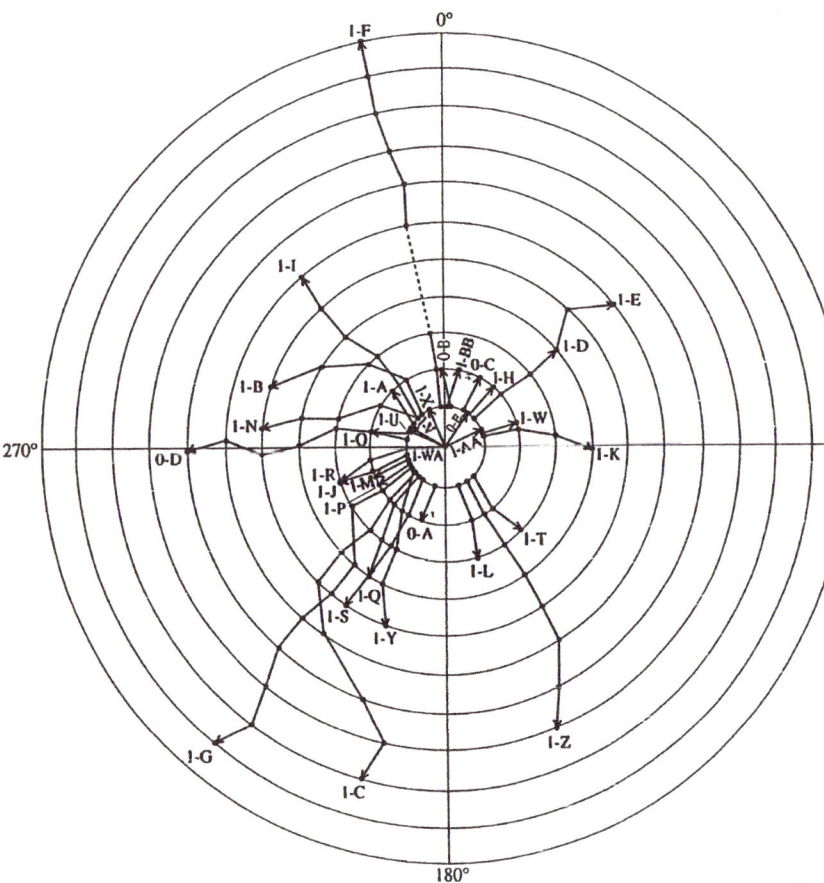

Figure 4-17.
A travel-direction analysis of measured *Eubrontes* trackways in the Exhibit Center shows trackways oriented in all directions of the compass. Magnetic north is 0 degrees in this diagram. Most of the trackways travel in fairly straight lines, and there is no obvious evidence of "herding" behavior.

and grooves made by reptiles walking on very firm sediments, such as those exposed to the drying effects of the air. In a revealing experiment, Farlow and Galton examined the footprints made by a lesser rhea, a bipedal, three-toed, ostrich-like bird, after it walked across a nearly hardened surface of plaster of Paris. Similar to the markings supposedly made by partly floating dinosaurs, the rhea footmarks only consisted of impressions of the three claws. Furthermore, some imprints of the lateral toes of the bird's foot were shallow, linear grooves rather like those on the Rocky Hill trackways. Tiny mounds of sediment were not seen behind any of the rhea's claw marks, but such mounds might be more likely if made in sand, a more fluid material than semi-hard plaster. Like the shallow, faint *Eubrontes* tracks also found on the upper sandstone bed at the Park, it is not difficult to visualize the curious claw marks as being produced by smaller, lighter theropods stepping on nearly dry, mica-rich shoreline sand. At those times, the sand surface was firm and cohesive enough to allow only the pointed claws of the animals to leave their impressions in the sediment.

The shorelines of the Jurassic lakes attracted large numbers of dinosaurs. Trackway evidence indicates that individuals or small groups of carnivorous dinosaurs made repeated visits to the lake margins. Perhaps dinosaurs came to the shores of the lakes to obtain drinking water, just as modern animals gather at water holes in semi-arid parts of the world. Another factor that draws animals to lake shorelines is the pursuit of food. In Africa, when antelopes, zebras and other *herbivores* come to water holes to drink, they attract a host of carnivorous predators, such as lions, leopards and hyenas. Surprisingly, however, no herbivores are represented among the tracks found at Rocky Hill. The footmarks of the large herbivores of the time, notably *prosauropod* and *ornithischian* dinosaurs, are entirely absent in the rocks at the Park. Compared with the abundance of theropod footprints, herbivore tracks are very scarce in the Central Valley as a whole. It is possible that most of the large Early Jurassic herbivores lived in forested habitats, obtained their drinking water from streams, and only rarely ventured to lake shorelines. Animals living in forested habitats would leave few preserved tracks. The reason for the rarity of herbivore tracks in the Valley remains a mystery.

Figure 4-18. The upper sandstone bed contains about 50 unusual marks, consisting only of the impressions of the tips of three claws. The central toe is represented by a small, semicircular claw imprint followed by a broad, shallow depression made by the front pad of the digit. Parallel grooves scratched by the claws of the side toes lie behind and on either side of the central toe impression. Some imprints of the side toes have tiny mounds of sediment behind the groove made by the claw. In Figure 4-16 these claw marks are drawn as groups of three circles arranged in a triangular pattern.

Since there is no evidence of any bones or tracks of Jurassic herbivores at the Park, what did the carnivorous dinosaurs that frequented the lakeshores eat? One very likely possibility might be that they ate fishes. Just as is the case today, many fishes resided in shallow water environments, where abundant sunlight and oxygenated waters enhanced the growth of plant and invertebrate populations, their food sources. During the monsoonal dry season, when lake levels dropped dramatically, large numbers of fishes may have become stranded along shoreline areas, attracting a variety of carnivores. The abundant smaller theropods, such as the animals that produced *Anchisauripus* and *Grallator* footprints, might also have been prey for *Dilophosaurus*-size animals. In the rainy season, when lake levels were high and fishes were not so easy to catch, perhaps theropods sought out food in more upland regions. Although the details of early Mesozoic terrestrial communities and *ecosystems* in the Valley are still unclear, evidence of thriving populations of plants, herbivores and carnivores is preserved in the rock record, and new discoveries steadily add to our understanding. The rocks and fossils at Dinosaur State Park provide a valuable window into the Mesozoic world.

Figure 4-19. Paleontologists disagree as to the origin of the claw marks shown in Figure 4-18. Walter Coombs Jr. first described these enigmatic impressions in a 1980 journal article, and interpreted them as "swimming tracks" made by a partially submerged theropod. Some scientists have concluded that the markings were more likely made by reptiles walking on very firm sediments. Others have suggested that these trace fossils are "underprints" or poorly preserved, largely eroded footprints made before most of the typical *Eubrontes* tracks were imprinted.

CHAPTER 5

Evidence:
Life of the Ancient Landscapes

The Mesozoic rocks of Connecticut and Massachusetts reveal evidence of the life and landscapes of Late Triassic and Early Jurassic times.

Although very few Late Triassic fossils have been found in the Central Valley, paleontologists nevertheless have a good idea of what plants and animals lived here, based on comparisons with organisms found in rocks of similar age in New Jersey, the American Southwest and elsewhere. The banks of the streams and rivers on the Valley floor were populated by a variety of plants, including forests of conifers, as well as ferns, *cycadeoids* and *horsetails* (Figure 5-1). The plant life supported numerous large herbivores and other organisms, including invertebrates (mollusks, worms, insects) and small vertebrates, such as fishes, amphibians and plant-eating reptiles. Small carnivorous dinosaurs and crocodile-like reptiles patrolled the riverbanks and sand flats in search of prey. Noticeably absent from this early Mesozoic scene, however, were flowering plants, birds and most mammals, for they had not yet evolved.

In spite of the diverse life that must have thrived in the Valley during the Late Triassic, the local rocks rarely preserve organisms from this time period. Only a handful of reptile skeletal remains have been discovered in Triassic-age strata in Connecticut, and with only minor exceptions, these fossils consist of bony fragments. The Triassic rocks in the Valley are all part of the New Haven Formation, which is composed mostly of red-brown deposits laid down in stream channels and on floodplains. Plants and animals that died and were buried in these oxygen-rich sediments and soils were rapidly broken down by decomposers, leaving a very sparse fossil record. Even the tracks of animals were not preserved in the shifting sands of the Late Triassic rivers and streams.

A Moment in the Triassic: The Sillin Mural and Late Triassic Diorama

In 1990, the Friends of Dinosaur Park Association commissioned William Sillin, a noted landscape artist, to create a large-scale mural of the Valley and its likely inhabitants in the Late Triassic. Because the local climate of that time was semi-arid, Sillin studied and sketched desert landscapes in Arizona, California and Nevada before starting the mural. The artist also consulted with specialists in the fields of geology and paleontology so as to make his creation factually accurate and realistic, down to the smallest detail. The mural took three years to complete. Soon thereafter, a life-size diorama featuring typical Late Triassic organisms was constructed in the foreground of the painting. As with the mural, experts were consulted so that the plant and animal models shown in the diorama were true to life. Together, the mural and diorama (Figure 5-2) vividly portray a moment of time in central Connecticut during the Late Triassic, more than 200 million years ago.

In the background of the mural, the crisp, low-angle light just before sunset brings out the colors and shadows on the well-worn hills and gullies of the ancient highlands to the east of the Valley. During the very brief rainy season, erosion of these exposed rocks supplied much sediment for the Valley. At the base of the highlands, coarse gravels were deposited on broad alluvial fans by the flash floods of temporary streams; finer sediments such as sand, silt and mud were carried to the toes of the fans and splayed onto the Valley floor. Hidden by the alluvial fan sediments lies the line of active faults that helped create the Valley. From time to time, downward movement on these faults produced powerful earthquakes that shook the Valley and its Mesozoic inhabitants.

A wide, shallow, *braided river* dominates the landscape of the near-flat Valley floor. Lowland rivers deposited sediments as a maze of bars, small islands and sand flats in and alongside a network of channels. Although the primary direction of river flow was to the southwest, some rivers flowed west and northwest away from the broad alluvial fans. The abundance of plant life on the banks and islands of this river suggests that it probably flowed year-round, although the amount of water it carried must have fluctuated greatly from the dry season to the brief rainy season. The exposed sand and mud flats in the foreground are covered by current ripples and *rills* which indicate a local flow direction to the northwest. The fallen tree trunks became lined up with the direction of water flow during a time of flooding on the river. The silvery, rain-slicked, puddled surfaces of the mud flats in the right foreground are the legacy of a recent thunderstorm, whose brightly lit clouds are now departing over the distant eastern horizon.

Figure 5-1. Giant horsetails along a Late Triassic riverbank.

Few plants were able to survive on the steep, arid slopes of the highlands, but the banks, bars and channel islands on the Valley floor were colonized by a variety of plant life. Tall conifer trees, such as *Pagiophyllum,* dominated Late Triassic forests in the Valley, but ginkgo trees also thrived. Cycadeoids (cycad-like plants), with their distinctive crown of leathery, fern-like leaves and thick, scaly trunks (Figure 7-2), also grew well along the rivers, and some of them reached the size of small trees. The damp sediments of the riverbanks were a hospitable environment for a variety of moisture-loving plants, especially ferns and horsetails. Horsetails, named for their resemblance to the braided tail of a horse, have green, cylindrical, jointed stems (Figures 5-1 and 8-6). Tiny branches and small, bristle-like leaves sprout in a *whorl* from the joints (known as nodes) on the hollow stems. A Late Triassic horsetail known as *Neocalamites* grew to more than six feet in height.

The Late Triassic reptiles and amphibians depicted in the mural include dinosaurs, but dinosaurs were not yet dominant. Crocodile-like phytosaurs were among the top predators during this time.

PLANTS

1 *Neocalamites*, giant horsetail
2 *Equisetites*, horsetail
3 *Selaginellites*, club moss
4 *Clathropteris*, fern
5 *Ginkgo*
6 *Ishnophyton*, cycadeoid
7 *Macrotaeniopteris*, cycadeoid
8 *Cladophlebis*, fern
9 *Pagiophyllum*, conifer
10 *Sanmiguelia*, (primitive flowering plant?)

ANIMALS

A *Metoposaurus*, amphibian
B *Hypsognathus*, reptile
C *Stegomus*, reptile
D Sphenodontid, lizard-like reptile
E *Rutidon*, phytosaur
F *Erpetosuchus*, crocodile
G *Coelophysis*, meat-eating dinosaur
H *Silesaurus*-like reptile

36

Figure 5-2. One of the highlights of the Exhibit Center at Dinosaur State Park is this striking mural and diorama depicting the life and landscapes of central Connecticut during the Late Triassic. The mural, by noted landscape artist William Sillin, measures 29 by 14 feet.

Figure 5-3. Key to selected plants and animals in Figure 5-2.

Figure 5-4. The Late Triassic phytosaur *Rutiodon* grew up to 12 feet in length. This powerfully built carnivore had short, sturdy legs and rows of bony armor plates on its back and tail. With its long jaws and sharp, conical teeth, *Rutiodon* likely preyed on fishes and land animals that ventured too near the water.

The general appearance of phytosaurs was remarkably similar to modern crocodiles, but with some important differences. These include the structure of the ankle joint, the position of the nostrils (farther back on the snout of phytosaurs), and the lack of an extensive bony palate that is present in crocodiles. Remains of a Late Triassic reptile similar to a phytosaur named *Rutiodon* (Figure 5-4) have been found in red sandstone of the New Haven Formation in Simsbury, Connecticut. The streamlined body, webbed feet and flat-sided tail indicate that *Rutiodon* was a good swimmer, but it was probably also capable of short bursts of speed on land. Like modern crocodiles, *Rutiodon* may have ambushed some of its prey by resting motionless in the river, with only its eyes and nostrils above water.

A distant relative of modern salamanders and newts, the huge amphibian *Metoposaurus* grew up to ten feet long (Figure 5-5). This animal possessed a highly flattened body, a stubby tail, and short, weak limbs. Metoposaurs never ventured far from the rivers and lakeshores that they inhabited, and they likely spent much time resting on the bottom in shallow water. They had enormous,

Figure 5-5. The 10-foot-long amphibian *Metoposaurus* had wide, massive jaws. It may have lured fishes into its gaping mouth by wiggling a worm-like projection attached to its tongue.

thick, flat skulls, with eye openings close to the nose, which enabled them to lie mostly submerged, crocodile-style, keeping an eye out for food as well as potential threats. It is thought that the diet of *Metoposaurus* consisted mostly of fishes, but it may have preyed on smaller amphibians and reptiles and invertebrates. Because this animal did not have strong legs or a powerful tail useful for sprinting after prey, it probably ambushed its victims. Although no fossils of metoposaurs have yet been found in the Central Valley, specimens are known from Late Triassic rocks in New Jersey, Pennsylvania and Nova Scotia.

Another common predator was the theropod dinosaur *Coelophysis* (Figure 5-2). *Coelophysis* was a sleek, lightly built, bipedal dinosaur about eight feet long and four feet tall at the hip. It had hollow limb bones and an average adult weight of 50 pounds. Its long, curved neck was balanced by a long, stout tail, which was held in the air and served as a counterweight. The hind limbs were powerful, rather like those of the modern ostrich. This dinosaur must have been a swift runner. *Coelophysis* walked on three toes equipped with large, curved claws. A fourth toe, known as the *hallux*, was short and higher up on the foot, seldom touching the ground. The front limbs were well adapted for grasping prey. The large hands had three long, clawed fingers and a very short fourth digit. The jaws could be opened widely and held some 50 curved and blade-like teeth, with fine serrations on the front and rear edges, ideal for tearing flesh. *Coelophysis* was an agile and effective hunter (Figure 5-6), but like *Rutiodon*, it could also be a scavenger, making a meal of already dead and decomposing creatures. No definite fossils of *Coelophysis* have yet been found in Connecticut, but it is well

known from Late Triassic rocks in the American Southwest, and it is very likely that it (or close relatives) inhabited the Valley. Hundreds of skeletons of this animal have been discovered at the Ghost Ranch site in New Mexico (Figure 5-7), and *Coelophysis* has been named the New Mexico state fossil.

Figure 5-6. This closeup from the Sillin mural shows *Coelophysis* about to make a meal of a sphenodontid, a small, lizard-like reptile.

Figure 5-7. Cast of one of the New Mexico *Coelophysis* skeletons on display in the Exhibit Center. Tiny bones inside the rib cage are the remains of the dinosaur's last meal. It was once widely believed that *Coelophysis* was cannibalistic, eating the young of its own kind. However, a re-examination of pertinent Ghost Ranch specimens reveals their stomach contents to consist of small, crocodile-like reptiles.

Erpetosuchus, a slender, crocodile-like carnivore with unusually long legs, was another resident of the riverbank. *Erpetosuchus* was a lightly built animal some two feet long, and presumably was a fast runner, using either two or all four of its long limbs. A partial skull of *Erpetosuchus* was discovered in 1995 along a roadcut in red sandy mudstone of the New Haven Formation near Cheshire, Connecticut (Figure 5-8). *Erpetosuchus* is thought to be closely related to *Terrestrisuchus*, whose fossil remains have been found in Late Triassic rocks in Wales.

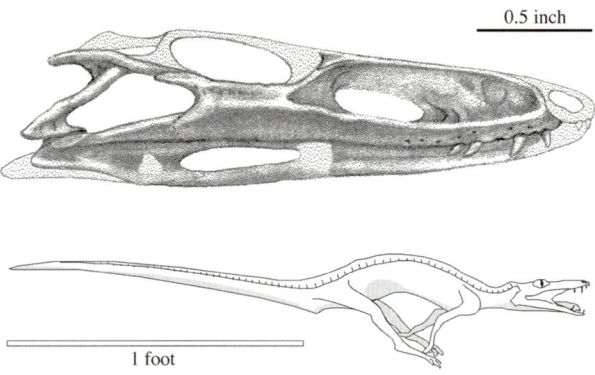

0.5 inch

1 foot

Figure 5-8.
Top: Reconstruction of the skull of *Erpetosuchus*, a crocodile-like reptile, based on a fossil specimen from Cheshire, Connecticut.
Bottom: Idealized sketch of *Erpetosuchus* running on all four limbs.

Plant-eating reptiles similar to *Silesaurus*, a dinosaur-like herbivore from the Late Triassic of Poland, are also represented in the Triassic mural. These animals had bird-like feet with three large toes, and were equipped with beaks suited for shredding vegetation. Slender and swift, they were some three feet long, had front legs almost as long as their hind legs, and were *quadrupedal*, running mostly on all four limbs. No fossils of these reptiles have yet been found in the Central Valley, but their presence is strongly suggested by abundant Late Triassic footprints found in New Jersey and other areas.

The heavily armored reptile *Stegomus* had an elongated, crocodile-like body and grew to a length of about six feet (Figure 5-9). Its full coat of body armor consisted of narrow, overlapping bony plates (scutes), which ran lengthwise in rows from neck to tail, covering the back. A separate series of rectangular scutes covered the belly and underside of the tail. This animal had a relatively small head and a pig-like snout; its blunt, peg-like teeth indicate a largely vegetarian diet. *Stegomus* may have used the tip of its snout and spade-shaped

lower jaw to dig in the soil for roots, *tubers* or fern *rhizomes*, and perhaps it occasionally ate worms, grubs or other soil insects. The only fossil of *Stegomus* known from the Central Valley is a sandstone cast of the dorsal body armor of the animal, discovered in a quarry near New Haven, Connecticut, in the 1890s.

Fossil remains of a diminutive, lizard-like reptile (Figure 5-10) were discovered by two school-children in 1967, in a stone wall in Meriden, Connecticut. Preserved in a loose block of red sandstone from the New Haven Formation was one of the most complete *Hypsognathus* skulls then known, along with much of the front part of the body. The short, broad skull, jaw structure and molar-like cheek teeth of *Hypsognathus* were suited to grinding tough, high-fiber plant matter, but the animal likely had a powerful bite, capable of crushing invertebrates with hard outer skeletons, such as crayfish and cockroaches. Burrows probably made by crayfish are abundant in the upper part of the New Haven Formation.

Figure 5-9. The armored reptile *Stegomus* grew to a length of six feet, and may have used its upturned, pig-like snout to probe the soil for plant roots and other food.

Figure 5-10. Reconstructions of *Hypsognathus* from the Late Triassic diorama. For protection, this foot-long reptile had bony spikes on its cheeks and a bony fri that covered part of its neck.

Jurassic Valley: The *Dilophosaurus* Diorama and Mural

About 200 million years ago, in the earliest Jurassic, further tilting and deepening of the Central Valley floor, coupled with a more humid climate, created long, narrow and deep lakes in the eastern half of the rift basin. The region was still under the influence of a monsoonal climate with two dominant seasons: a rainy season and a dry season. Because much of the Valley floor was flat or gently sloping to the east, even small variations in the amount of rainfall could bring about dramatic changes in lake size. During wet intervals, rising lake waters would flood shoreline, beach and low-lying areas, depositing mud, silt and sand; in dry periods, shrinking lakes would expose lake-margin sediments to the baking heat of the sun. Dinosaurs left their tracks in the muds and wet sands of lake shorelines and river floodplains, and these tracks were covered and preserved by fresh sediment when lake or stream levels rose again. Dinosaur tracks are very abundant in the silty mudstones and sandstones that once were lake-edge or floodplain deposits; in fact, it is unusual not to find tracks in this kind of rock.

A highlight of the Park's Exhibit Center is the mural and diorama depicting the shoreline of an Early Jurassic lake in the Central Valley (Figure 5-12). In the foreground, the broad expanse of bare, exposed sand and mud flats covered with dinosaur tracks indicates that steady heat and evaporation have led to a drop in lake levels. Flashes of lightning from the impending thunderstorm in the distance, however, show that the rainy season is on its way.

Two robust bipedal dinosaurs patrol the shoreline in the center of the exhibit, leaving behind their tracks as the only evidence of their passing. These predatory carnivores are reconstructions of a theropod dinosaur called *Dilophosaurus* (Figure 5-13). Although no bony remains of *Dilophosaurus* have yet been discovered in the Central Valley, skeletons of this dinosaur have been found in Early Jurassic

strata in Arizona (Figure 5-14), in rocks roughly the same age as those at the Park. Bones that match the abundant large *Eubrontes* tracks at the Park have never been found in the Valley. However, of all the known contemporary dinosaurs, *Dilophosaurus* is considered by paleontologists to be the best candidate for the maker of these tracks, based on its size, foot structure and stride length.

At 20 feet in length, nearly six feet tall at the hip, and weighing up to 1,000 pounds, *Dilophosaurus* was one of the largest carnivores on Earth during Early Jurassic times. The enlarged, elongated head of this dinosaur holds more than 50 sharp, saw-edged teeth that grew to three inches in length, ideal tools for grasping prey, and slicing and tearing flesh (Figure 5-11). However, the front part of the upper jaw is partially separated from the rest of the upper jaw by a peculiar "subnaral gap," and this somewhat delicate jaw construction has led some paleontologists to suggest that this dinosaur was more of a scavenger than an active predator. Other researchers, noting the long teeth at the front of the jaws and the position of the nostrils far back on the snout, have theorized that this animal ate fishes. The sturdy, powerful legs and arms of *Dilophosaurus* were equipped with long, sharp, curved claws, which must have been useful weapons (Figure 5-15). Likewise, a swat from this dinosaur's muscular tail could easily knock smaller animals off their feet. At one locality in Arizona, several fossil specimens of this reptile were found close together, suggesting that this dinosaur hunted in packs.

Figure 5-11. The curved, serrated, three-inch teeth of *Dilophosaurus* served well for grasping prey and tearing flesh. The colorful, twin bony crests on top of the head were probably used for display.

Figure 5-12. The Exhibit Center mural and diorama depicting a forested lake shoreline in Connecticut during Early Jurassic times. Dominating the foreground are two reconstructions of a formidable predatory dinosaur known as *Dilophosaurus*.

Figure 5-13. The *Dilophosaurus* model that forms the centerpiece of the Jurassic diorama is the first life-size reconstruction of this dinosaur. At 20 feet in length and weighing up to 1,000 pounds, *Dilophosaurus* was one of the largest Early Jurassic carnivores.

Figure 5-14. A nearly complete skeleton of *Dilophosaurus* from rocks in Arizona that are approximately the same age as those in the Park. The upward arching of the neck and tail bones is quite common in dinosaur and fossil bird skeletons. Scientists debate whether this posture results from an agonized death, such as by asphyxiation, or from post-mortem muscle contractions.

Figure 5-15. Striding on two powerful, muscular legs, *Dilophosaurus* must have been a fast runner. It walked on its three large toes, each tipped with a stout, sharp claw. A fourth toe, high on the rear of the foot, rarely made contact with the ground. The fleshy pads on its feet were covered by scaly skin.

Figure 5-16. A turkey-size, herbivorous ornithischian dinosaur within a cluster of tall horsetails. As with nearly all models and paintings of dinosaurs, the actual skin coloration pattern can only be theorized. Many dinosaurs may have had some pattern of spots or stripes on their backs for camouflage.

Within a patch of tall horsetails at left-center of the diorama are three turkey-size Early Jurassic ornithischian dinosaurs with distinctly bird-like beaks and feet (Figure 5-16). These fast, slim herbivores had arms that were much shorter than their legs, and they could walk or run on either two legs or using all four limbs. When running on two legs, their long tail would counterbalance the front half of their body. No skeletal remains of ornithischian dinosaurs have yet been unearthed in the Mesozoic rocks of the Central Valley, but their presence is strongly indicated by fossil tracks that have been given the name *Anomoepus*. These tracks likely were made by a dinosaur similar to *Lesothosaurus*, a primitive ornithischian from the Early Jurassic rocks of southern Africa.

At the far end of the diorama, a trio of long-necked prosauropod dinosaurs is partly concealed among the cycadeoids and ferns growing densely on the forest floor. Prosauropods were a diverse group of plant-eating dinosaurs that lived during Triassic and Early Jurassic times. The prosauropod *Anchisaurus* is featured in the diorama and mural (Figure 5-17).

Figure 5-17. The prosauropod dinosaur *Anchisaurus* grew to more than eight feet in length. Standing on its hind legs, it could feed on the leaves of tree ferns, tall cycadeoids and shrub-size young conifers, a food source not available to the shorter, smaller herbivores of the Jurassic forest. It ate small pebbles, which were used in its gizzard to shred tough, fibrous plant matter.

The name *Dilophosaurus* means "two-crested lizard." The function of the twin bony crests atop the skull of this dinosaur has been a subject of some debate among paleontologists. The semicircular crests, constructed of thin bone, were too fragile to be used for protection or defense. The crests contain no blood vessels, so it is unlikely they were used by this animal to regulate its temperature. It is very possible that the delicate crests were brightly colored, much like the red, fleshy "comb" on a rooster's head, and so could be used in threat displays to ward off enemies or rivals in disputes over territory. It has also been theorized that the crests were used for species recognition or to attract potential mates.

Figure 5-18. Foot bones of *Anchisaurus*, discovered in a loose red sandstone boulder from a gravel pit in Ellington, Connecticut. The largest digit is five inches long.

This dinosaur is known from several skeletons discovered in the Portland Formation near Manchester, Connecticut, and elsewhere. *Anchisaurus* was a lightly built dinosaur with a small head, a slender, elongated neck, and a long, slim body and tail. Adults averaged 1.5 feet high at the hip, grew to more than eight feet in length, and weighed an estimated 60 pounds. *Anchisaurus* typically walked on all four legs, but its strong hind limb and ankle bones enabled it to walk bipedally as well.

Unlike the four-toed theropods, the hind foot (*pes*) of *Anchisaurus* had five digits. Four of the toes were large, with numerous small pads and blunt claws (Figure 5-18). The narrow hand (*manus*) of this animal also had five digits, with the fourth and fifth digits being smaller and clawless. The thumb was equipped with a large curved claw, which might have been used to pull down tall vegetation, to dig up edible roots, or as a defensive weapon. Unfortunately, the hand and foot structure of *Anchisaurus* cannot be matched up with any known tracks from the Central Valley, but the very large tracks called *Otozoum* are thought to have been made by a much larger prosauropod relative.

Historically, the most productive fossil bone locality in the Central Valley was the old Wolcott Quarry in the Buckland section of Manchester, Connecticut (Figure 5-19). During quarrying operations for brownstone in the 1880s, three well-preserved prosauropod skeletons were discovered, along with portions of two other individuals. The *Anchisaurus* bones and other skeletal remains at the Wolcott Quarry are preserved because they were quickly covered by sediment before they had a chance to deteriorate. The Quarry is located only some three miles from the faulted eastern edge of the Valley, and the red-brown sandstone at the site may represent deposits accumulating near the base of an alluvial fan, in an environment of rapid deposition that preserved bone material. Unfortunately, much of the Wolcott Quarry was blasted and bulldozed to make way for access roads and parking lots for the Buckland Hills Mall.

Gliding in the air at one end of the diorama and perched on a conifer branch at the other end are

Figure 5-19. Thick red-brown sandstone beds in a portion of the old Wolcott Quarry, Manchester, Connecticut. Several well-preserved prosauropod skeletons were discovered here when the Quarry was in operation.

Figure 5-20. Reconstruction of the pterosaur *Dimorphodon*, a carnivorous flying reptile. *Dimorphodon* had an oversize head with deep and wide tooth-filled jaws, and a four-foot wingspan. Its long, thin, pointed tail was tipped by a diamond-shaped flap of skin used for flight control.

models of a pterosaur known as *Dimorphodon* (Figure 5-20). Pterosaurs were neither dinosaurs nor birds, but rather carnivorous flying reptiles. The head of this active aerial predator most closely resembles that of the modern puffin, a fish-eating bird. The teeth and jaws of *Dimorphodon* suggest that it actually did eat fishes, which were abundant in the Early Jurassic lakes of the Central Valley. One individual in the diorama is chasing a dragonfly, and it is possible that this animal's diet also included the larger insects in the forests, such as the grasshoppers and giant cockroaches seen in the exhibit. Fossils of *Dimorphodon* have not been found locally, but skeletons from Early Jurassic rocks in England are well known.

"Fossil" Lakes of the Jurassic

Much of what paleontologists know about the Early Jurassic life of the Central Valley is derived from fossils preserved in ancient freshwater lake deposits. Some of the best specimens of fossil plants, fishes and invertebrates from Early Jurassic lake deposits in the region can be seen in the Exhibit Center display cases at the Park. Just like modern lakes, these ancient lakes were ecosystems composed of complex, interacting communities and populations of micro-organisms, plants and animals. Bacteria and algae formed the base of most food chains in the Jurassic lakes. Geochemists have identified algae-derived organic matter in samples of black shale from the region, and large *oncolites* (rounded structures made by algae) have been found in the East Berlin Formation near Durham, Connecticut (Figure 5-21). Pollen, spores and fragments of land plants blown or washed into the lakes during storms added to the producer base of aquatic food chains (Figures 5-22 and 7-3). A variety of *zooplankton* must have existed in the lakes, but fossil evidence of these microscopic organisms is lacking. The shoreline waters supported a host of larger invertebrates, such as insects, worms, mollusks and *crustaceans*. Some gray and red shale units contain abundant burrows, crawling traces and invertebrate trails.

Figure 5-21. Oncolites from the East Berlin Formation near Durham, Connecticut. Microbes living in warm, sunlit lake waters precipitated concentric layers of limestone around a carbon-rich core, probably once a tree branch. The left-hand specimen is four inches long.

Figure 5-22. A three-inch-long fragment of conifer foliage from the Portland Formation, South Hadley Falls, Massachusetts.

Among the most abundant invertebrate residents in the near-shore regions of these lakes were small, bivalved crustaceans called conchostracans and ostracods, creatures that still thrive today. Conchostracans are tiny shrimp-like animals that live inside a protective, two-part, clam-like shell (Figure 5-23). Ostracod fossils from the Valley resemble tiny hot dog rolls (Figure 5-26), and can rarely be seen without a microscope. The hard outer skeletons of these *arthropods*, once thought to be rare in the Valley, have now been discovered at nearly every locality that produces fossil fishes. These crustaceans likely were a major part of the diet of some of the fishes that inhabited the lakes.

Bivalve mollusks also thrived in the shallow waters of the Early Jurassic lakes (Figures 5-24 and 5-35). Mollusk fossils collected from the Central Valley rocks are nearly indistinguishable from the modern *Unio*, a clam found in many freshwater habitats. The present-day *Unio* is a plankton-filtering bivalve that lives exposed on lake and stream-channel floors, or it can burrow a short distance into the sediment. Gray sandstone beds of the Portland Formation in Suffield, Connecticut, contain detailed external and internal molds of clams as well as numerous burrows excavated by the mollusks. The association of the bivalves with their burrows leads to the conclusion that these clams were preserved where they actually lived.

Figure 5-23. Conchostracan shells from the Portland Formation, Chicopee Falls, Massachusetts. Conchostracans, commonly called "clam shrimp," are tiny arthropods that produce a protective, two-part outer skeleton or shell. These specimens average 0.25 inches in length.

Insects must have been abundant in the Early Jurassic forests of the Central Valley, but their delicate remains were rarely preserved. The only existing insect fossils (other than crawling traces or burrows) obtained from local rocks were collected from lake deposits. The first specimens of Mesozoic insects discovered in North America were aquatic larvae (*Mormolucoides articulatus*), obtained in the 1850s from poorly documented localities on the banks of the Connecticut River near Gill, Massachusetts. In October 1991, the "lost" *Mormolucoides* beds were rediscovered by Phillip Huber and the author in Montague, Massachusetts. Excavations at the site have produced plant fossils, some 5,000 larvae, and the remains of beetles (Figure 5-25). The latest studies suggest that *Mormolucoides* is the larval stage of a beetle. Well-preserved adult insect remains, including the elytra (stiff wing covers) of beetles (Figure 5-26), and a tiny, calcite-replaced wing of a cockroach (Figure 5-27), were collected in the 1980s from a thin bed of chocolate-brown *lacustrine* claystone in the Portland Formation near Suffield, Connecticut.

Figure 5-24. A freshwater clam (Unio) from the Portland Formation, Suffield, Connecticut. Concentric growth lines are visible in this internal mold of the shell. The specimen is 1.7 inches long.

Figure 5-25. Insect larvae (*Mormolucoides articulatus*) collected from Montague, Massachusetts, in 1991. The largest larvae are about an inch long and 0.25 inches wide, and, including the head, are composed of 13 segments. They are preserved as silvery, carbonized impressions on black shale. A) a mass-mortality layer; B) a complete specimen; C) close-up of head region of B, showing mandibles (jaws).

Figure 5-26. A beetle elytron (stiff wing cover or forewing) from the Portland Formation, Suffield, Connecticut, showing remnants of its dark-lined vein pattern. The elytron is 0.38 inches long. Several tiny, sausage-shaped ostracods are visible above the wing.

Figure 5-27. A remarkably preserved cockroach wing from the Portland Formation, Suffield, Connecticut. The wing is 0.25 inches long and is covered by a thin layer of white calcite, which preserved the intricate vein structure.

47

Figure 5-28. A four-inch *Semionotus* from the East Berlin Formation near Durham, Connecticut, on display in the Exhibit Center. *Semionotus* can be distinguished from the other Central Valley fishes by the stout scales (fulcra) on the leading edge of its fins, and by the relatively smooth surface of its skull bones.

Figure 5-29. Skull details on a large *Semionotus* from the Shuttle Meadow Formation, North Guilford, Connecticut. Note the short jaws with peg-like teeth, the eye socket and the smooth skull bones. Including the bones covering the gills, the skull is two inches long.

The Central Valley has long been recognized as one of the premier regions in the country for the collection and study of Early Jurassic fishes. Fossil fishes from Westfield, Connecticut, only a few miles from the Park, were described in scientific publications as early as 1816. More than 10,000 specimens from localities throughout the Valley have found their way into museums. In Connecticut and southern Massachusetts, fish fossils are abundant in the *microlaminated* black shales of the Shuttle Meadow, East Berlin and Portland Formations. These black shales were formed from layers of black mud that accumulated in the deeper, anoxic portions of perennial lakes. Fishes were certainly present in the rivers and streams of the Central Valley during the Triassic Period; the complete absence of fossil fish remains in the red rocks of the Triassic New Haven Formation reflects a lack of preservation.

At first glance, most fossil fishes from the Early Jurassic rocks of the Valley closely resemble modern fishes in their shape, arrangement of fins, and overall appearance and structure. However, the scales and skull bones of the Jurassic fishes are very thick and heavy compared with most modern forms, and they are coated with a glossy, enamel-like calcified tissue called ganoine. Park visitors often ask if the fossil fish specimens in the Exhibit Center have been varnished or lacquered. The luster on the scales, fins and bones is completely natural, and the fossils have not been "treated" in any way. Most of the specimens on display were prepared by "airbrasion," a technique similar to sandblasting, but using sodium bicarbonate (which is harder than the shale but softer than the fish parts) as the abrasive.

The fishes residing in the large Central Valley lakes lived in the upper, oxygen-rich waters or close to shoreline areas where food was plentiful. Typically, after death, the bodies of modern fishes bloat with decomposition gases and float at the surface where their remains can be eaten or can disintegrate. It is likely that many of the heavy-scaled Jurassic fishes quickly sank into deeper water where few scavengers and decomposers existed. Dead fishes may have lain at the bottom of lakes for extended periods. Over time, accumulations of fine mud and clay covered, flattened and preserved their bony skeletons. At several black-shale localities, the occurrence of "fish-kill layers" (zones in which hundreds of fish fossils are found) indicates sudden mass mortality of fish populations, perhaps caused by the shrinking of lakes during the dry season, or poisoning of the water by volcanic eruptions.

Four *genera* of fossil fishes are currently recognized from the Early Jurassic rocks of Connecticut. The most common *genus* is known as *Semionotus* (Figures 5-28 and 5-29). Semionotids have been recovered from all of the known fossil fish sites in the Valley, and have a worldwide distribution in early Mesozoic strata. *Semionotus* was a heavily built fish; the largest local specimens are 16 inches in length. *Semionotus* had a short mouth armed with peg-like or conical teeth; its small mouth and mobile upper jaws suggest a browsing or nibbling method of feeding, and it probably also could engulf prey by suction. Its diet may have included conchostracans and other crustaceans, small mollusks, and soft-bodied invertebrates. Semionotid fishes from the Central Valley display an astounding variety of body forms and shapes, ranging from slender types to deep-bodied, hunchbacked varieties and everything in between. More than three dozen species of *Semionotus* have been described from early Mesozoic rocks in the Valley and nearby New Jersey. Several exemplary specimens of *Semionotus* are on display at Dinosaur State Park.

Figure 5-30. A six-inch specimen of *Redfieldius* from the Shuttle Meadow Formation, North Guilford, Connecticut. Distinguishing features of this genus include the skull bone ornamentation, the extreme rearward position of the dorsal fin, and the tiny scales on the front edges of the fins.

Figure 5-31. Reconstruction of *Ptycholepis* from the Early Jurassic rocks of the Central Valley. This six-inch-long fish is recognized by its narrow, ridged and grooved body scales and similar ridge and groove ornamentation on the skull bones.

Redfieldius is a genus of fossil fish first discovered in a stream exposure in Middlefield, Connecticut, in 1836. Since then it has been found in considerable numbers in all of the Jurassic sedimentary formations in Connecticut, and most recently in the Turners Falls area in Massachusetts, on the banks of the Connecticut River. Specimens of *Redfieldius* average eight inches in length from the snout to the tip of the caudal (tail) fin (Figure 5-30). The snout of *Redfieldius* is covered by tiny conical *tubercles*, and resembles a pincushion. It has been suggested that these tubercles supported a large, fleshy upper lip, and that *Redfieldius* was a bottom feeder, scooping up its food with its upper lip and elongated lower jaw.

In the Central Valley rocks, *Ptycholepis* is a relatively rare form; it has been found at only a handful of localities in the Shuttle Meadow and East Berlin Formations. *Ptycholepis* was a slender, streamlined

Figure 5-32. *Ptycholepis* from the Shuttle Meadow Formation, North Guilford, Connecticut, with an "exploded head," likely due to the buildup of decomposition gases, which burst the skull and scattered the head bones.

fish, with a torpedo-shaped body and delicate fins. Its length from the snout to the tip of the caudal fin averages slightly over six inches. The large mouth was armed with numerous small teeth and could be opened widely. *Ptycholepis* was presumably a fast swimmer and an active predator. A reconstruction of *Ptycholepis* can be seen in Figure 5-31. A very unusual specimen of *Ptycholepis* from North Guilford, Connecticut, is shown in Figure 5-32. The body of this *Ptycholepis* is nearly intact, but the skull bones have been separated and scattered, as if the head has exploded. While this fish lay slowly deteriorating at the bottom of a lake, it is likely that cavities within the head filled with decomposition gases, and the buildup of these gases literally burst the skull apart. Similar "gas bursts" have been observed in other fossil fish specimens, particularly in the belly region.

The *coelacanth* fish *Diplurus longicaudatus* (Figures 5-33 and 5-35) is the largest but least common of the Central Valley fossil fishes. Fewer than 30 specimens of this sturdy, lobe-finned fish have been collected from the Shuttle Meadow and East Berlin Formations of Connecticut, and from the Turners Falls area of Massachusetts. At up to three feet long, *Diplurus* was a top predator in local Early Jurassic lake ecosystems, with a diet largely consisting of other fishes. Its mouth could be opened widely, and it would gulp its victims whole, grinding them up using its bony palate and throat plates. Coelacanths were once thought to have been extinct since the *Cretaceous Period*, but a fish remarkably similar to *Diplurus* was dredged from the depths of the Indian Ocean off South Africa in 1938. Coelacanth specimens are now known from the African coast to Indonesia. Often called "living fossils," coelacanths are considered to be the closest links between fishes and the early amphibians that emerged from the seas to colonize the land in mid-Paleozoic times.

Figure 5-33. The coelacanth fish *Diplurus longicaudatus* from Turners Falls, Massachusetts. This near-complete specimen is 26 inches long and lacks only the front portion of the head. *Diplurus* can be identified by its two fan-shaped dorsal fins, its distinctive caudal fin (at rear), its well-preserved internal skeleton, and other features.

Fossilized fecal remains known as "coprolites" are common in many of the black-shale deposits of the Valley. Coprolites have been observed at nearly all of the localities that produce fossil fishes, and most are likely derived from fishes. Typically they are structureless masses of black, carbonaceous material, but occasionally the larger coprolites contain visible fish scales and bones. Chemical analyses of the large coprolites reveal high levels of phosphate, similar to levels found in fish bones. The size and composition of the phosphatic coprolites (Figure 5-34) suggests that they came from *Diplurus*, the predatory coelacanth. At least one specimen of *Diplurus* from North Guilford, Connecticut, actually contains one of these phosphatic coprolites within its body. A number of *Semionotus* and *Redfieldius* specimens from the same locality also contain round, knobby or grainy coprolites about the size of rabbit droppings. Coprolites that can be positively attributed to dinosaurs or other reptiles have not been found in local rocks.

Very little evidence of land life has been found in the black shales that formed in the deeper portions of Early Jurassic perennial lakes. In spite of the many thousands of fossil fishes obtained from beds of microlaminated black shale (Figure 5-36), only a handful of isolated reptile teeth have been found

Figure 5-34. This three-inch coprolite (fossilized fecal matter) has a high phosphate content and likely was produced by *Diplurus*, the fish-eating coelacanth. Central Valley coprolites vary in shape from round or ovoid to elongated, tapering masses, and range in size from a fraction of an inch to more than five inches long.

in perennial lake deposits in the Valley. The best of these finds, a tooth of a theropod dinosaur (Figure 5-37), is on display in the Park Exhibit Center. The tooth is three-quarters of an inch in length, conical and gently curved, with fine serrations on the posterior margin. Although frogs, salamanders, turtles, lizards, sphenodontids and other vertebrates were present on Earth at the time when the great lakes of the Central Valley existed, their remains have yet to be discovered in any local Early Jurassic deposits.

Figure 5-35. In a dramatic scene from the Exhibit Center's "underwater" diorama and mural, the predatory *Diplurus* stalks two redfieldiids. On the lake floor, several clams are shown in their natural filter-feeding position, burrowed into the sediment with about half of the shell exposed.

Figure 5-37. This 0.75-inch tooth from a carnivorous theropod dinosaur was found in 1970 by Bruce Cornet and the author during excavations at the North Guilford locality. It is on display in the Exhibit Center, housed in a plexiglass box with a top that magnifies the specimen.

Figure 5-36. Robert Demicco and the author (right) in a quarry pit in the Shuttle Meadow Formation near North Guilford, Connecticut, in August 1976. Numerous fossil fishes, coprolites, plant remains and isolated reptile teeth were obtained from the fossiliferous "deep lake" black shales at this site.

CHAPTER 6

Traces:
Fossil Footprints of the Central Valley

Tracks as Fossils

The tracks and trails left behind by animals are part of a special category of fossils known as trace fossils,

so named because they preserve only evidence of an animal's activity and behavior, rather than preserving its actual body parts. Sedimentary rocks in the Central Valley sometimes display a variety of traces of Mesozoic organisms: swimming and crawling traces, resting traces, burrows, boreholes and tunneling activity, and evidence of feeding behavior. These traces are often the only record that the animals existed. Tracks and other trace fossils in the Valley typically occur in rocks that were once lakeshore muds and sands or river floodplain deposits. The raindrop impressions, ripple marks and mud cracks (Figures 3-9, 6-1 and 6-12) frequently associated with the tracks and traces verify their origin in very shallow water, or in sediments exposed to the drying effects of the air. Because they are only imprints in sediment, tracks and trails (Figure 6-2) have a greater chance of being preserved in the fossil record than the body parts of organisms, which can be destroyed by scavengers and decomposers.

Figure 6-1. Wind-driven waves or currents created these shallow-water ripple marks on this slab of gray-tan siltstone from the East Berlin Formation, West Springfield, Massachusetts. The slab measures 10 by 12 inches.

Figure 6-2. This slab of red mudstone from the Portland Formation, Suffield, Connecticut, displays a seven-inch-long crawling trace of an invertebrate, probably a crustacean or a large insect. As the animal wriggled its way from left to right through the silty mud, tiny mounds and ridges of sediment were piled up on both sides of the trace.

The type and quality of any trace fossil left by an animal depends on its body structure, particularly the anatomy of its legs or feet, and also on the particle size and firmness of the sediment (Figure 6-3). Given their body weight, their prominent toes and claws, and the large pads on the underside of their scaly feet, the Early Jurassic dinosaurs of the Central Valley were well equipped to produce excellent tracks. Some tracks are very complete and detailed (Figure 6-4), but the majority of tracks found in the local rocks are not well preserved because the sediment was either too hard or too soft (Figure 6-5). Tracks made in soft sand or mud begin to fill in immediately once the foot has been lifted, and, when preserved, these *penetrative tracks* may not even display the outline of the dinosaur's toes. Footprints made on hardened mud or gravelly surfaces are also shallow and indistinct. Scaly skin impressions are very rare even under ideal conditions because, during normal walking, feet tend to slip and slide in muddy or sandy sediment.

Figure 6-3. Modern bird tracks from a partly-evaporated mud puddle, East Berlin, Connecticut. The footprint at far right was made in firm mud, and clearly shows the bird's toes and the fleshy pads on the underside of its foot. The tracks become increasingly less distinct as the bird stepped into softer sediment. Note the developing mud cracks, created as the sediment dries out and shrinks. The footprints are three inches long.

Figure 6-4. A six-inch negative theropod track from the Portland Formation in Cromwell, Connecticut. This footprint was made under nearly ideal conditions of sediment particle size and consistency, allowing preservation of foot-pad and claw details.

55

Figure 6-5. Poorly preserved theropod tracks, Portland Formation, Bloomfield, Connecticut. These footmarks, made in very soft gray mud, sank deeply into the sediment, and were mostly filled in with mud as the trackmaker's feet were lifted. The largest track is five inches long.

Tracks known as "underprints" or "overprints" are vague, shallow, poor-quality traces made in sediment layers an inch to several inches below or above the actual surface layer the animal walked upon. A significant number of Central Valley dinosaur tracks contained in the collections at Amherst College, Wesleyan, and Yale University are of this type (Figure 6-6). A few tracks at the Park are also underprints. Not recognizing that many underprints and overprints are simply deformed versions of dinosaur surface tracks, many past researchers have considered them to be the tracks of a variety of reptiles, amphibians and even mammals. Some local track beds have been so heavily trampled by dinosaurs that the normally horizontal sediment layers and the tracks turn into an unrecognizable jumble. This condition is popularly called "dinoturbation" by paleontologists. Like any other fossils, tracks can be deformed, damaged or destroyed during the years it takes for the sediment to harden into rock, or at a later time by tectonic forces.

The abundance of dinosaur tracks in local rocks suggests that the alternating wet and dry (monsoonal) climate of the Central Valley during the Early Jurassic was a vital factor in their preservation. As shown in the foreground of the *Dilophosaurus* diorama, many tracks were made in semi-hardened mud or clay-rich sand along the shorelines of large lakes. During dry periods,

it would not take long for these track impressions to harden in the baking tropical heat. With the return of steady rains, lake levels would again rise, overflowing lake margins exposed during the dry season, and covering previous shorelines and their tracks with fresh layers of sediment. Year after year, the seasonal shrinking and expansion of Early Jurassic lakes buried and preserved thousands and thousands of tracks in the sedimentary rocks of the Valley.

Steady erosion of the Central Valley since Mesozoic times has uncovered many footprint-bearing layers, especially in stream channels and along the banks of rivers. Countless more tracks have been brought to light in quarrying operations for building and paving stone, and in excavations for roadways and building foundations, such as at Dinosaur Park. In the 19th century, a number of small quarries were opened solely for the purpose of collecting tracks and larger trackways. Horse transport was used to bring massive slabs back to Amherst, Yale, Dartmouth, Harvard, Wesleyan and other colleges, universities and museums. Rocks containing footprints often can be most easily split along the layers with the tracks, revealing not only the actual indented imprint the animal made, but also producing a natural cast of the sediments that filled in the original track impression. As noted earlier, indented track imprints are called "positives," and the raised "out-prints" are known as "negatives" (Figure 4-13).

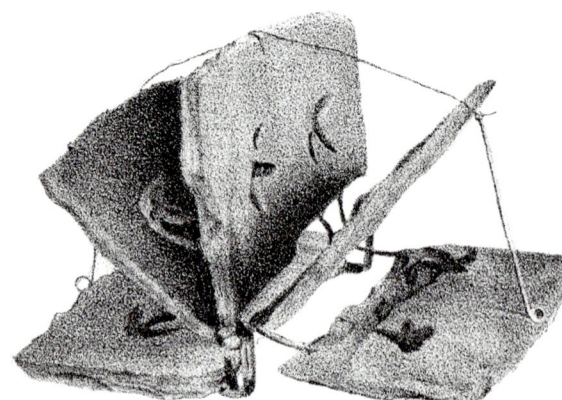

Figure 6-6. An illustration of underprints and overprints, showing the same two tracks passing through five successive layers of sandstone, each an inch thick. Edward Hitchcock called this specimen the "fossil volume," and attached the rock layers together with metal hinges to form a "stony book."

The tracks now visible on the bedrock layers at Dinosaur Park are entirely positive impressions made by dinosaurs walking on the sandy shore of a Jurassic lake. Numerous slabs of sandstone containing negative prints were removed from above the main trackway layer; some of these slabs can be found on Park grounds, but others were donated to colleges, schools and museums around the State. The rock layers containing the trackways are not entirely composed of sand, but also contain thin zones of finer sediments, such as silt and clay. The upper surface of the main trackway layer is one of these zones, and it is also coated with *muscovite* mica, a very flaky mineral abundant in some sediments. This combination of clay and mica created a zone of weakness between the main trackway layer and the rock beds immediately above, and, during digging, the rock frequently split at this weak zone. Flakes of mica, acting like a natural Teflon coating, allowed many positive and negative prints to separate cleanly and perfectly. The Park trackways thus owe their existence to a precise set of circumstances: sediment of the proper grain size and consistency, burial and preservation of the tracks by later deposits of sediment, zones of clay and mica, and, of course, the dinosaurs that visited the ancient lake shorelines.

Hitchcock and the Origins of Ichnology

History does not document when fossil footprints were first discovered in the Early Jurassic sedimentary rocks of the Central Valley. A popular legend records their earliest notice by European settlers in 1802, when a teenager named Pliny Moody unearthed a sandstone slab covered with tracks while plowing on his family farm in South Hadley, Massachusetts (Figure 6-7). The scientific importance of Moody's slab was not immediately recognized, and it was put to practical use as a doorstep. The slab and its impressions, which looked like they had been made by enormous birds, attracted the curiosity of neighbors. Not yet aware of the existence of dinosaurs, local residents commonly referred to the footprints as "turkey tracks" or those of "Noah's raven," with reference to the biblical flood. Though it was not known at the time, these imprints were the first evidence of dinosaurs found in the Western Hemisphere. The slab was purchased in 1839 by Amherst College, where it is now prominently displayed in the College's Museum of Natural History.

Figure 6-7. Engraving of the footprint quarry at the Moody farm, South Hadley, Massachusetts, as it looked in the 1850s. According to Hitchcock, the earliest discovery of fossil footmarks in the Central Valley was made in 1802 by Pliny Moody when his plow unearthed a sandstone slab containing tracks.

In the spring of 1835, several residents of Greenfield, Massachusetts, noticed more of these "stony bird tracks" on locally quarried rock slabs being laid down for a new sidewalk. James Deane, a successful Greenfield surgeon with interests in natural science, brought the tracks to the attention of the Rev. Edward Hitchcock, a minister, scientist and educator at Amherst College. Intrigued by the geological and biological implications of the footmarks, Hitchcock devoted the last three decades of his life to the study of Central Valley trackways and trails. He published more than 30 reports on footmarks, and built up a collection at Amherst that exceeded 20,000 impressions, the largest accumulation of trace fossils in the world (Figure 6-8). Two of Hitchcock's most important books are on display at the Park: his 1858 *Ichnology of New England*, and the 1865 supplement to that volume.

At first, many in the greater scientific community questioned the authenticity of Hitchcock's "ornithichnites" (or footmarks of birds), and doubted whether they were actually footprints. His interpretations of their origin were validated in the autumn of 1840 when members of a committee of eminent geologists visited his localities and examined his evidence. To his final days, in spite of mounting evidence to the contrary, Hitchcock remained convinced that many of his three-toed tracks were those of large, flightless birds, not unlike the extinct moa of New Zealand or the modern ostrich. The Central Valley tracks are now known to be primarily those of reptiles, including small to medium-size dinosaurs and crocodile-like forms (Figure 6-22). However, in light of the modern view that birds are the direct descendants of theropod dinosaurs, Hitchcock's conclusions, made more than a century ago, now appear closer

Figure 6-8. This brownstone slab, probably from Portland, Connecticut, measures three by five feet and displays more than 50 well-preserved theropod footprints. Track-side-down, it was used as a paving stone in the streets of Middletown, Connecticut, for 60 years. Hitchcock called this specimen the "gem" of his collection.

to fact than fiction. History has come to recognize Hitchcock's role as the founder of the field of ichnology, the study of fossil tracks.

Common Fossil Footprints in the Central Valley

In his series of publications from 1836 to 1865, Edward Hitchcock named close to 50 *ichnogenera* of Central Valley vertebrate footprints and more than 100 *ichnospecies*, not including the names he created and later discarded. Unfortunately, most of Hitchcock's track names are not valid when modern standards of ichnology are applied. Many of Hitchcock's earlier specimens were of relatively poor quality: shallow or infilled tracks, partially eroded traces, overprints, underprints and penetrative tracks, and "superimposed" prints, in which animals had stepped twice in the same place. These imperfect tracks, though, were often the ones described and illustrated by Hitchcock in his many publications; they became the *type specimens* for his ichnogenera and ichnospecies. Hitchcock's footprint descriptions emphasized measurements of the length of toes, pads and other parts. Frequently, footprints that varied only in size were given different names, rather than being interpreted as evidence of older or younger individuals of the same type of animal.

In the Early Jurassic rocks of the Central Valley, poorly preserved vertebrate tracks are far more abundant than well-preserved footprints. However, the only tracks that can be identified, classified and named with any confidence are those that show details of foot structure, such as well-defined toes, joints, foot pads, and claw and skin impressions. Modern studies have reduced the number of valid ichnogenera to fewer than ten, although a greater variety of different animals certainly could have

made similar-looking tracks. The most abundant tracks in the Valley are the three-toed impressions made by small to large carnivorous theropod dinosaurs; these tracks have been given the names *Eubrontes*, *Anchisauripus* and *Grallator*. Theropod tracks typically are *tridactyl* prints, with three narrow to moderately wide toes and distinct claws. Another toe, the hallux, was short and located higher on the foot, and contacted the ground only when the animal's feet sank into very wet, muddy sediment. Some theropods likely had a tiny *vestigial* fifth toe, but no mention of this toe appears in any descriptions of Central Valley theropod footprints.

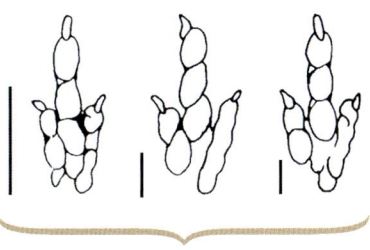 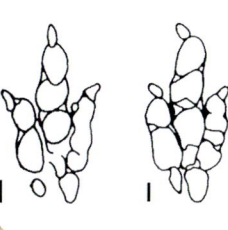  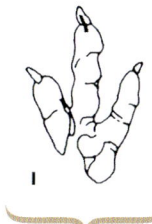

Grallator   *Anchisauripus*   *Eubrontes*

Eubrontes, *Anchisauripus* and *Grallator* differ from one another mainly in terms of size: overall length, width, and how far the central toe extends beyond the side toes. This has led many researchers to conclude that they were made by different types of theropod dinosaurs. However, Paul Olsen of Columbia University remeasured hundreds of tracks at Yale and Amherst, and observed a nearly continuous change in footprint shapes with increasing size (Figure 6-9). He concluded that *Eubrontes*, *Anchisauripus* and *Grallator* tracks could have been produced by individuals of different ages, perhaps even fully grown adults, younger adults and juveniles of the same dinosaur species. Olsen suggested that the size and shape variations seen in Central Valley theropod tracks could be a result of allometric growth, in which some parts of the foot changed shape as the animals grew. Variations in footprint shape also can be caused by flexing of the foot during contact

Figure 6-9. Diagram of typical theropod footprints from Early Jurassic rocks in the Central Valley and elsewhere, arranged from left to right in order of increasing size. Scale bars equal two centimeters. The track types show a near-continuous change in shape with increasing size; larger tracks have relatively shorter middle toes. The various tracks may have been made by different species of theropods or by younger or older individuals of the same dinosaur species. Herein, the names *Grallator*, *Anchisauripus* and *Eubrontes* are used in a general way to refer to small, medium-size and large theropod tracks.

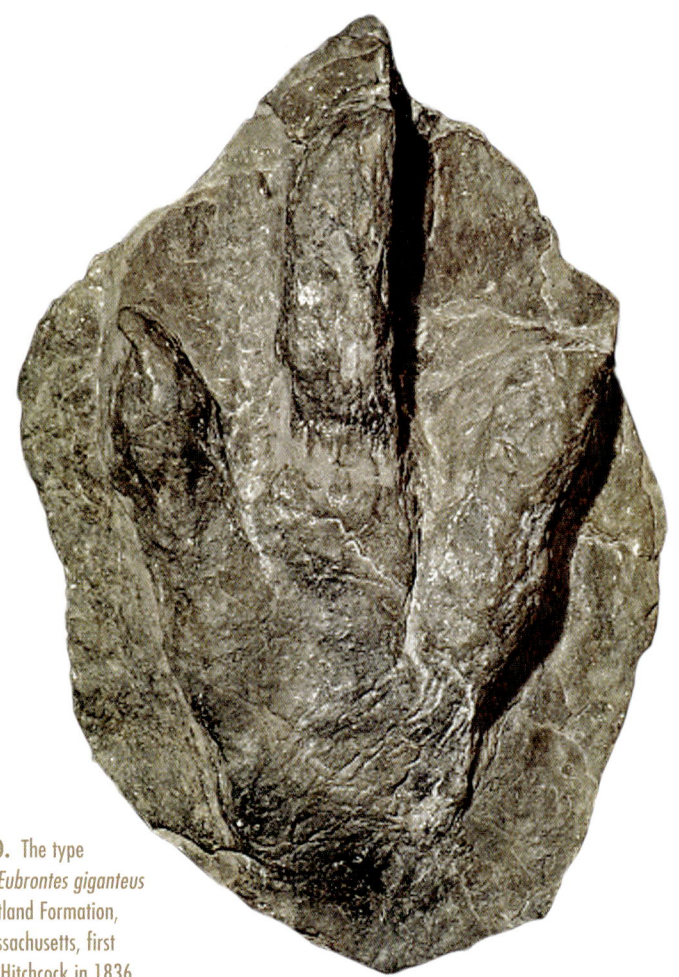

Figure 6-10. The type specimen of *Eubrontes giganteus* from the Portland Formation, Holyoke, Massachusetts, first described by Hitchcock in 1836. The central toe of *Eubrontes* is relatively short compared with the surrounding toes, though on average it extends four inches beyond the side toes. Track length is about 15 inches.

Eubrontes giganteus tracks likely were made by a theropod dinosaur similar to *Dilophosaurus*, an Early Jurassic carnivore originally discovered in Arizona. Since 1991, *Eubrontes* has been recognized as the state fossil of Connecticut.

The ichnogenus *Grallator* ("stilt walker") was created in 1858 by Hitchcock for some of his *Eubrontes*-like footprints that were very much smaller and more slender. *Grallator* tracks are tridactyl traces up to six inches in length (Figure 6-12). It has been suggested that *Grallator* footprints were made by a fast, agile Early Jurassic theropod about the same size, shape and weight as *Podokesaurus*, known from a partial skeleton discovered in 1910 in South Hadley, Massachusetts. *Podokesaurus* closely resembles *Coelophysis*, the well-known theropod from the Late Triassic of the American Southwest. At the Park, some 20 well-preserved *Grallator* tracks can be seen on a slab of red-brown sandstone in the Exhibit Center.

with the sediment, or later deformation of the sediment or rock. Emma Rainforth of Ramapo College also concluded that *Eubrontes*, *Anchisauripus* and *Grallator* footprints are size variants within the same ichnogenus, but other workers emphasize that further research is necessary to resolve these nomenclature issues. The names *Eubrontes*, *Anchisauripus* and *Grallator* are therefore used herein in a general way to refer to large, medium-size and small three-toed theropod tracks.

Hundreds of large theropod tracks can be seen at Dinosaur Park, both in the Exhibit Center and on the grounds. In 1836, Edward Hitchcock assigned the name *Ornithichnites giganteus* to similar footprints from what is now called the Dinosaur Footprint Reservation in Holyoke, Massachusetts. In 1845, he changed the ichnogenus to *Eubrontes*, a name translating to "true thunder," a possible reference to the impact of the massive track-makers' feet on the ground (Figure 6-10). Compared with other local theropod footprints, *Eubrontes* tracks are broad and large, with lengths ranging from 10 to nearly 18 inches (Figure 6-11).

Figure 6-11. A very large, well-preserved, negative *Eubrontes* from the Shuttle Meadow Formation, west of Higby Mountain, Middletown, Connecticut. Two of the toes show distinct claw imprint fillings. A 12-inch ruler gives scale.

All of the tracks on the rock are negative prints, formed as infillings of the original track impressions. Most of the footprints are between four and five inches in length, with very distinct casts of the foot pads. Fine, worm-like trails of invertebrates are very abundant on the surface of this slab.

The ichnogenus *Anchisauripus* ("*Anchisaurus* foot") was named by R.S. Lull of Yale University in 1904. Lull mistakenly believed these tracks were made by *Anchisaurus*, a dinosaur then thought to be a theropod, but now understood to be a plant-eating prosauropod. Based on what he thought were common hallux impressions, he also concluded that the *Anchisauripus* track maker was *tetradactyl* and walked on four toes. Rainforth's studies indicate that the "hallux claw imprints" seen by Lull on most of his *Anchisauripus* specimens are actually either underprint claw impressions from other tracks, mud cracks, invertebrate traces, or questionable markings. As currently understood, *Anchisauripus* tracks are medium-size tridactyl traces some six to ten inches long, the foot being narrower than in *Eubrontes* tracks, but wider than in *Grallator* footprints (Figure 6-13). *Anchisauripus* is a "middle-size" theropod track, with size and shape characteristics in between those of the other two theropod ichnogenera common in the Valley. At the Park, a few footprints small enough to be called *Anchisauripus* are present on the bedrock exposures in the Exhibit Center, but they are not easily recognized among the abundant *Eubrontes* trackways.

In 1847, Edward Hitchcock chose the name *Otozoum* ("giant animal") for what he thought was "perhaps the largest and most extraordinary track yet brought to light in this valley." A trackway containing four of these new tracks was discovered on a 10-foot-long slab of gray-brown sandstone in South Hadley, Massachusetts (Figure 6-7). The collector was the same Pliny Moody who, according to Hitchcock, first recognized fossil footprints on the family property about 45 years earlier. The type specimen is preserved in the Museum of Natural History at Amherst College. Although they are relatively rare in the Early Jurassic strata of the

Figure 6-12. Well-preserved *Grallator* tracks on a mudcracked brownstone slab from the Portland Formation, Portland, Connecticut. Compared with the larger theropod tracks, the track width in *Grallator* is narrower, the toes are more parallel, and the central toe is longer when compared with the side toes. *Grallator* claw marks are relatively long, gently curved and quite pointed. The largest footprint measures five inches.

Figure 6-13. A seven-inch negative *Anchisauripus* footprint preserved in gray-tan silty sandstone from the Portland Formation, Cromwell, Connecticut. *Anchisauripus* tracks have size and shape characteristics in between those of *Eubrontes* and *Grallator*.

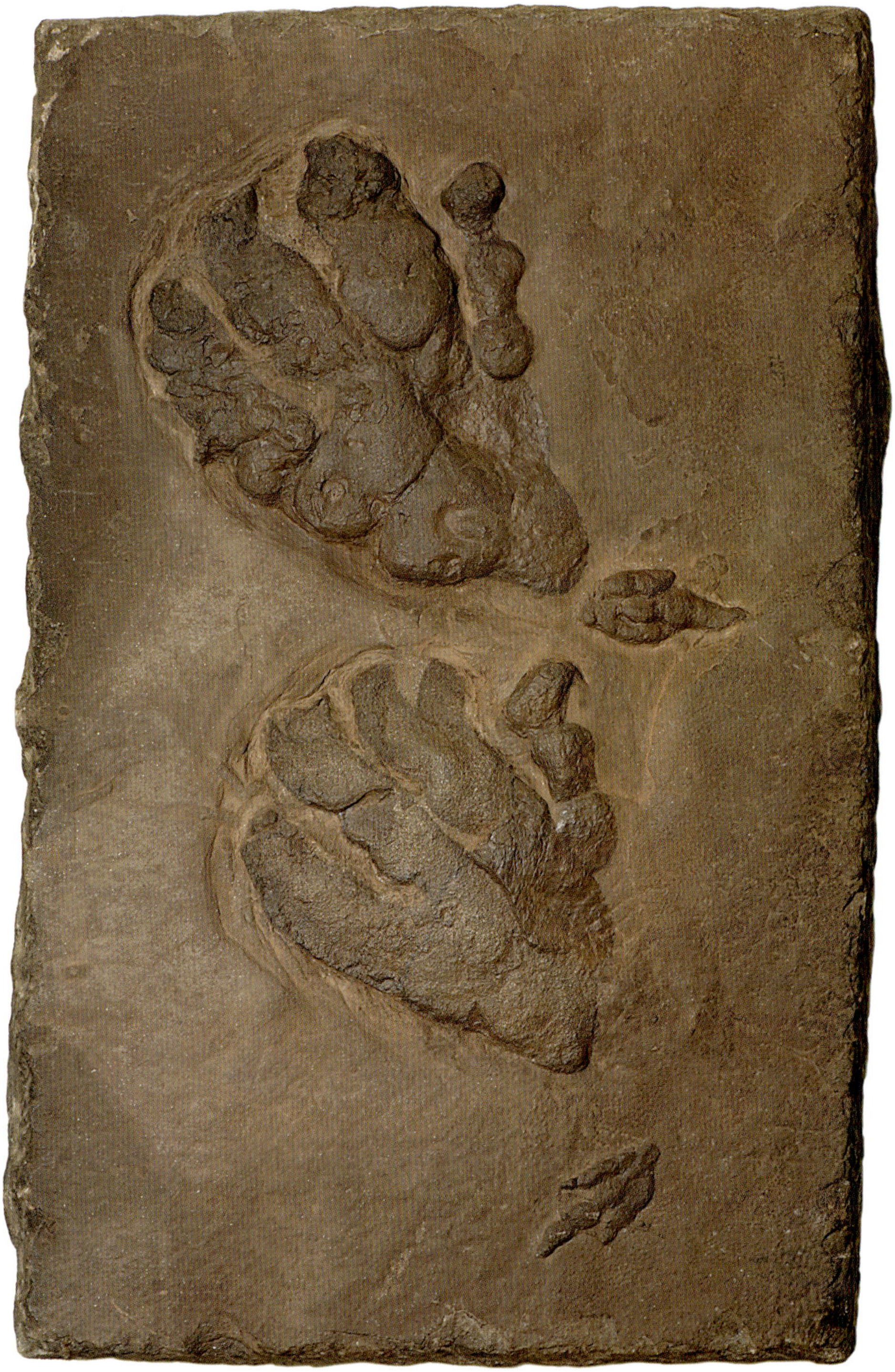

Figure 6-14. Footprints of *Otozoum*, the probable tracks of a prosauropod dinosaur, from the brownstone quarries at Portland, Connecticut. The largest *Otozoum* track is 17 inches long. Two small *Grallator* prints accompany the massive footmarks.

Central Valley, *Otozoum* footprints are truly impressive, reaching lengths of some 20 inches.

Well-preserved *Otozoum* tracks display four large, thick, nearly parallel toes with distinct pad impressions, as well as the trace of a short fifth toe pad (Figure 6-14). Claw marks on *Otozoum* tracks are often rounded, short and blunt, and are not easily seen on some specimens. Hitchcock theorized that *Otozoum* traces were made by a huge, bipedal, frog-like amphibian; later workers have suggested marsupial mammals, crocodile-like reptiles, and various types of dinosaurs. Comparisons of *Otozoum* footprints with the feet of prosauropod dinosaurs now suggest that the *Otozoum* track maker was a bipedal prosauropod that walked on its toes. *Otozoum* prints very likely were produced by a prosauropod roughly twice the size of *Anchisaurus* (Figure 5-17). Unfortunately, bony remains of this animal have not yet been found in local rocks.

One of the highlights of the Park's track collection is a massive block of brownstone from Portland, Connecticut, that displays an *Otozoum* trackway with five tracks in a row. The slab weighs more than a ton, and was quarried from Portland in 1896 at a time when brownstone was a highly valued building stone. The *Otozoum* tracks are very well preserved casts (negatives) from a thick sand layer deposited on top of the original track impressions; the largest footprint measures about 15 inches in length (Figure 6-15). All of the tracks show filled "mud crack" traces that seem to sprout from parts of the footprints. The force of the heavy dinosaur's footstep split the semi-hardened sediment, and the

Figure 6-16. Close-up of scaly skin impressions on the *Otozoum* footprint in Figure 6-15. Skin impressions, quite uncommon on local theropod tracks, are often preserved on *Otozoum* footprints, perhaps because of the size and weight of the track maker.

cracks were later filled in. The small, oval-to-circular depressions in front of several of the tracks likely are from lumps of sediment dropped as the dinosaur lifted its muddy feet off the ground. Parts of the surface of the slab show evidence of flowing water currents, from lower right to upper left on the rock. The preservation of fine skin details on the footprints (Figure 6-16) suggests that they were deep enough to be protected from erosion or were made later on nearly dry sediment. Tiny invertebrate crawling traces are visible at lower right on the slab, next to the last *Otozoum* track.

Anomoepus, an ichnogenus now considered to represent the tracks of a small, slender and swift ornithischian dinosaur, was first established by Edward Hitchcock in 1848. The *Anomoepus* track maker was mainly bipedal, but occasionally walked, rested or sat with all four of its limbs touching the ground (Figure 5-16). The ichnogenus name translates to "unlike foot," a reference to the marked difference in structure of the foot (or pes) impressions compared with the manus (Latin for "hand") imprints (Figure 6-17). The largest *Anomoepus* pes tracks are upward of eight inches long, though most are smaller. The track maker had four toes on its foot, but imprints of the hallux are usually seen only in sitting tracks. Three-toed pes tracks are thus typical. At the Park, a three-foot-long slab of brownstone from the Portland quarries contains some 15 *Anomoepus* tracks. The tracks are negatives, and appear to be bipedal walking tracks, as only pes traces are clearly visible.

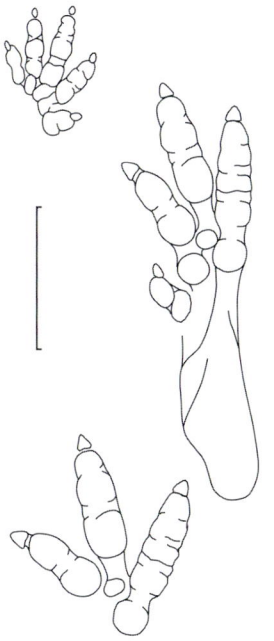

Figure 6-17. Drawings of *Anomoepus* tracks, probably the footprints of an ornithischian dinosaur. From bottom to top: a typical pes (foot) track made while the animal was walking; a "sitting" pes track; and a manus (forefoot) track. *Anomoepus* pes walking traces show three widely spread toes, and the rear pad of the outside toe lines up with the central toe, an unusual and diagnostic feature. Sitting *Anomoepus* pes traces often include the hallux, impressions of the heel bones (long, oval markings behind the toes), and imprints of the tail. Manus impressions show five digits spreading forward and outward in a fan-shaped arrangement. Scale bar equals four inches.

The tracks look like they were made at the same time, and nearly all are headed in the same direction, suggesting that the track makers were traveling in a group. Another *Anomoepus* trackway on display shows three successive positive footprints, accompanied by a tail-drag trace.

The overall shape and structure of most *Anomoepus* footprints is very bird-like, and, to his final days, Hitchcock believed that his *Anomoepus* tracks were created by birds (Figure 6-18). However, Hitchcock was troubled that the manus tracks of *Anomoepus* were not bird-like, and more closely resembled the forefeet of lizards and some mammals. He mistakenly concluded that the *Anomoepus* track maker did not walk on its front limbs, and thought that the fan-shaped arrangement of the fingers resembled the digits of an "expanded wing." The Amherst professor must have been reassured when he read descriptions of the newly discovered *Archaeopteryx* from Jurassic rocks in Germany. This primitive bird possessed a blend of bird-like and non-bird-like characteristics.

Figure 6-18. A superb walking trackway of *Anomoepus* from Gill, Massachusetts, part of the Hitchcock Collection at Amherst College. The ends of the slab are covered by raindrop impressions which slightly obscure the tracks; the central part of the slab is smooth. A shallow puddle must have covered the middle region at the time of the shower, preventing one track from pitting and eroding. Most of these tracks are five inches long.

Figure 6-20. Tiny, quadrupedal pes and manus tracks of *Batrachopus* in red shale from the Portland Formation, South Hadley, Massachusetts. These tracks lack imprints of the backward-pointing fifth toe.

Figure 6-19. Diagram of a *Batrachopus* trackway showing multiple pes and manus imprints, likely those of a primitive crocodile. *Batrachopus* pes (hind foot) impressions range up to nearly three inches in length and typically display four toes. Claw impressions on the toes are short and are not always present. Manus traces are smaller than the pes tracks, and have five toes, though typically only four are imprinted. Some toes of the manus point forward, some point sideways, and the fifth toe, when visible, points to the rear.

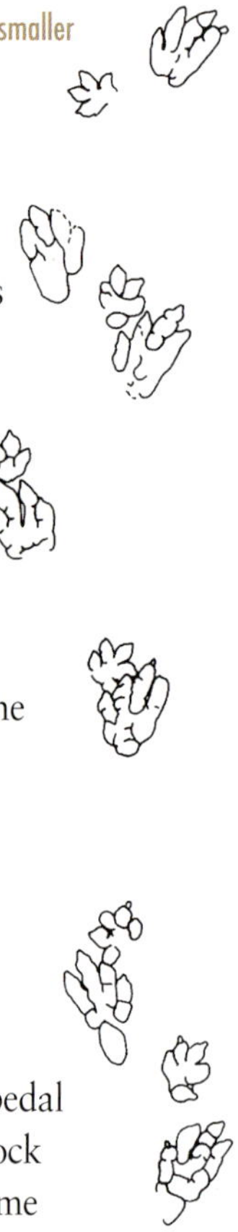

In the end, still expressing some cautious doubt, Hitchcock decided that the bird-like features of *Anomoepus* outweighed the evidence suggesting a different type of animal. Paleontologists now think that *Anomoepus* tracks were made by a plant-eating, turkey-size ornithischian similar to *Lesothosaurus*, known from the Early Jurassic of southern Africa.

Footprints assigned to the ichnogenus *Batrachopus* are the most common small, non-dinosaur tracks in the Central Valley, and they were the first footprints recognized to have been made by a habitually quadrupedal animal (Figures 6-19 and 6-20). Hitchcock named this ichnogenus in 1845. The name translates to "frog foot," and, as Hitchcock collected more and more of these tracks, he changed his opinion regarding the type of animal that made them: first a bird, then an amphibian, next a lizard-like creature. In 1858, he finally settled on an animal with a blend of marsupial mammal and crocodile characteristics. As was his custom, Hitchcock also changed the ichnogenus name four times to reflect the different animals he thought made the tracks. The striking similarity between *Batrachopus* footprints and the tracks of crocodiles was noted by James Deane in his last and most important work, published in 1861. In 1904, R.S. Lull suggested that *Batrachopus* was probably the trackway of *Stegomosuchus longipes*,

a slender, long-legged, armored reptile (now thought to be a primitive crocodile) known from a partial skeleton discovered in the Portland Formation near Longmeadow, Massachusetts. Recent workers have similarly concluded that the *Batrachopus* track maker was a small, fully terrestrial early crocodile.

One of the most curious specimens in the Park's track collection is a small slab of red sandy mudstone whose smooth surface bears odd scratch marks and small, quadrupedal tracks that have been called *Antipus flexiloquus*. This slab was found in the East Berlin Formation in Newington, Connecticut, by Bruce Cornet, and was donated to the Park. *Antipus* ("opposite foot") *flexiloquus* ("uncertain") was named by Edward Hitchcock in 1858. The ichnospecies name reflects Hitchcock's uncertainty as to whether this track was made by turtles, lizards or even salamanders. *Antipus* tracks are very scarce, and the few known examples do not show the structure of the pads on the foot and hand or other details (Figure 6-21). *Antipus* and *Batrachopus* tracks share many similarities, and some researchers have argued that *Antipus* tracks are just poorly preserved *Batrachopus* traces. However, the manus and pes prints of *Antipus* are far apart, whereas those of *Batrachopus* are close together and may overlap. Paleontologists are still in doubt as to which animal is responsible for the trackways of *Antipus flexiloquus*. Some have suggested crocodile-like creatures, others,

lizard-like animals related to the modern tuatara of New Zealand. Another hypothesis proposes that these trackways were made by a pterosaur, and if so, they represent the earliest pterosaur tracks from North America.

Figure 6-21. Quadrupedal tracks of the type specimen of *Antipus flexiloquus* from Gill, Massachusetts. Pes impressions of *Antipus* show four narrow toes and sometimes a partial fifth digit or heel mark. Manus traces typically show five widely splayed digits, often rotated outward from the trackway midline. The kind of animal that made *Antipus* tracks is still debated. The largest track on this slab is about one inch long.

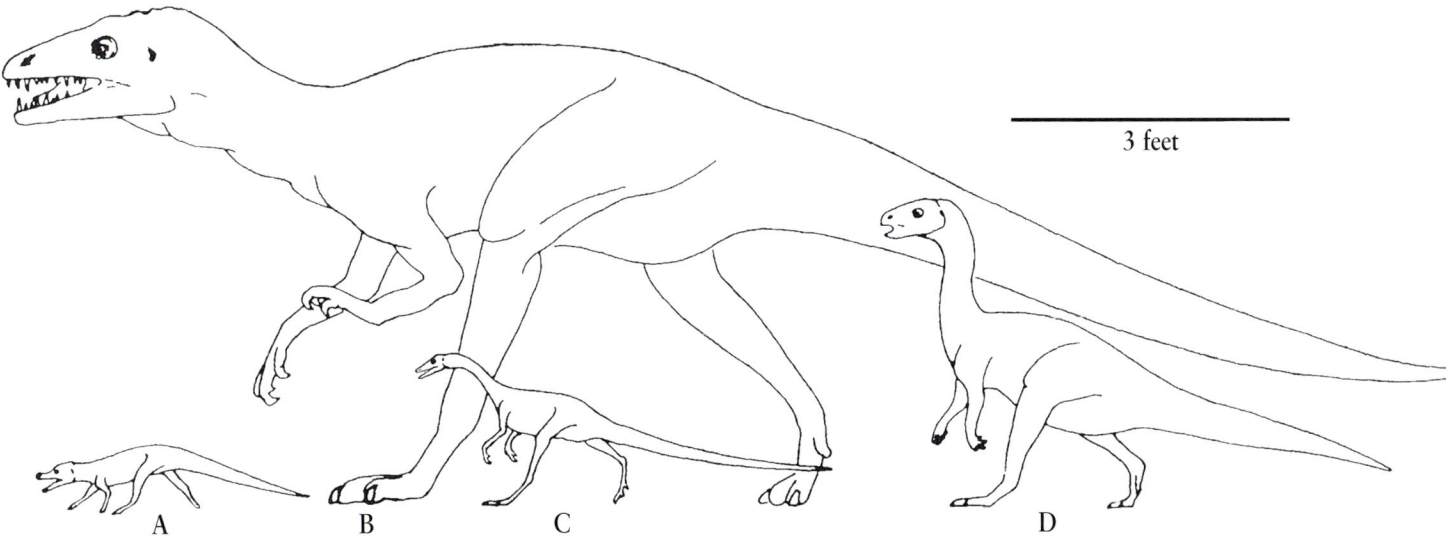

Figure 6-22. Possible appearance of some Central Valley Jurassic track makers: A) Small crocodile, producer of *Batrachopus* footprints; B) Large, carnivorous theropod dinosaur, responsible for *Eubrontes* and *Anchisauripus* tracks; C) Small theropod, maker of *Grallator* tracks; D) Ornithischian dinosaur, producer of *Anomoepus* footprints.

CHAPTER 7

The Arboretum of Evolution

An arboretum is a collection of trees and shrubs grown for exhibition or for study.

The "Arboretum of Evolution" at Dinosaur State Park was developed and designed for educational purposes, but it also creates an attractive park-like setting for the grounds surrounding the Exhibit Center (Figure 7-1). It is a living museum containing more than 200 species of conifers and flowering plants whose close ancestors can be traced back to the age of dinosaurs and before. The plants in the Arboretum were selected and arranged to display their evolutionary relationships and diversity. Even before Park visitors enter the Exhibit Center, they are exposed to a unique landscape of ancient plants, a setting that complements the Jurassic trackways and displays.

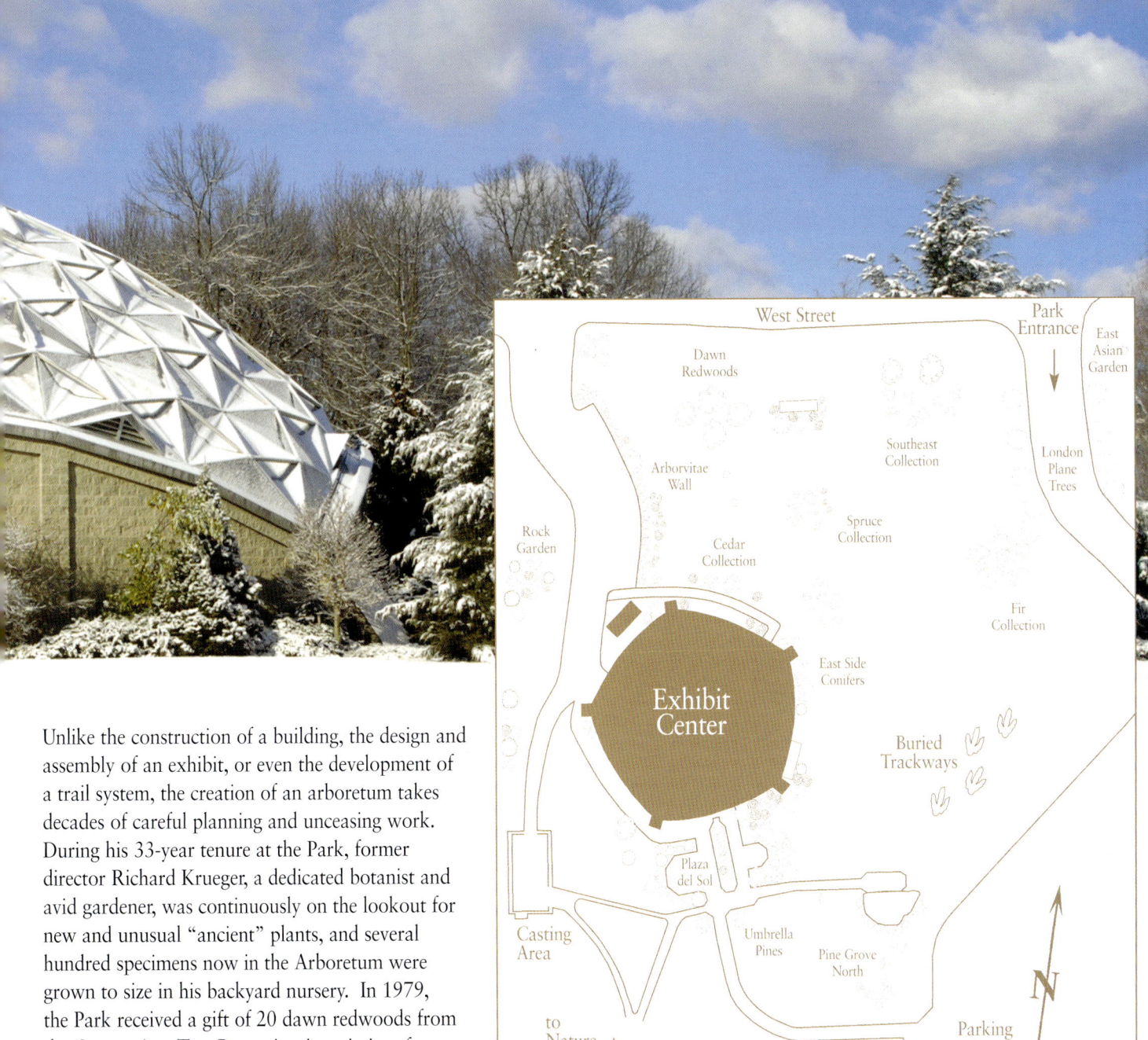

Figure 7-1. Location map for Arboretum plantings.

Unlike the construction of a building, the design and assembly of an exhibit, or even the development of a trail system, the creation of an arboretum takes decades of careful planning and unceasing work. During his 33-year tenure at the Park, former director Richard Krueger, a dedicated botanist and avid gardener, was continuously on the lookout for new and unusual "ancient" plants, and several hundred specimens now in the Arboretum were grown to size in his backyard nursery. In 1979, the Park received a gift of 20 dawn redwoods from the Connecticut Tree Protective Association; four of these now-mature trees border West Street, north of the Exhibit Center (Figure 7-17). In 1981, Ted Childs of Norfolk, Connecticut, made a generous donation of 32 young conifers, many of which are still in the collection. That year, with the help of the Hillside and Rocky Hill Garden Clubs, a small nursery was established on Park grounds, and many trees were grown from seeds, seedlings and rooted cuttings. Since 1990, the purchase of new plants for the Arboretum from catalogs and local nurseries has been made possible by a horticultural fund established by the Friends of Dinosaur Park. The Arboretum is a testament to Krueger's dedication, artistry and efforts, and to that of Park staff. A catalog of the Arboretum collection lists nearly 1,300 specimens, representing 57 plant families and more than 120 genera. After more than a quarter-century of development, the Arboretum of Evolution is reasonably complete and mature; it serves as a valuable instructive resource and greatly enhances the natural beauty of the Park grounds.

Figure 7-2. Reconstructions of Early Jurassic ferns and a cycadeoid (center) from the *Dilophosaurus* diorama in the Dinosaur State Park Exhibit Center. Most cycadeoids had stout, scaly trunks and a crown of fern-like fronds with stiff, leathery leaves. Some ferns and cycadeoids could grow to the height of small trees.

The Fossil Record of Land Plants

Plants first established a foothold on land more than 400 million years ago. The fossil record from that time includes simple bryophytes (related to modern mosses and liverworts) and the earliest *vascular plants*. The early land plants mainly reproduced by the dispersal of spores. Vast, dense forests of gigantic ferns and club mosses existed during the Carboniferous Period, or "coal age," when shallow swamps covered much of the land surface. During a time of widespread desert conditions at the end of the Paleozoic, many spore-producing plants were replaced by tough, drought-resistant, seed-bearing plants such as ginkgoes, cycadeoids and conifers. These seed plants would become the dominant vegetation in the early Mesozoic.

When dinosaurs first became numerous in the Late Triassic, nearly all of the major groups of vascular plants except the *angiosperms* (flowering plants) were in existence. Common plant fossils found in the early Mesozoic rocks of the Central Valley include conifers, cycadeoids, ferns and remains of large, tree-like horsetails (Figures 7-2 and 7-3). By the middle of the Jurassic Period, conifers had become more diverse, and most modern conifer families were well established. Conifers are particularly hardy plants and are well adapted to living in semi-arid climates or monsoonal climates with extended dry seasons, conditions that prevailed

Figure 7-3. This piece of black shale from the Shuttle Meadow Formation in Durham, Connecticut, displays a three-inch portion of a cycadeoid (*Otozamites*) frond as well as part of the stem of a horsetail (*Equisetites*) with parallel vein structure. The black, oval mass at lower right is a coprolite, likely produced by the predatory fish *Diplurus longicaudatus*.

in the Central Valley during the Early Jurassic (Figure 7-4). Conifer needles are covered with a thick waxy cuticle that prevents water loss. Compared with broad, flat leaves, needles have a smaller surface area exposed to drying sunlight and wind. The same adaptations that enable conifers to thrive in hot, dry conditions also allow them to survive in cold, dry northern climates, and also in upland or mountain habitats. The resin found in the leaves and wood of conifers does not freeze like the watery sap found in most flowering plants. Conifers reached their peak of diversity about 100 million years ago, and have been in slow decline ever since. Climate change and competition from the aggressive and prolific flowering plants have been the primary factors for their decline. Nevertheless, an impressive selection of scarce and unusual conifers is preserved and featured in the Park Arboretum.

Angiosperms, plants that produce flowers, are the most diverse and widespread plants on Earth today. Flowers are a highly effective means of reproduction, typically using wind or insects as a means of spreading their pollen. Unlike the ginkgoes and conifers, angiosperms have seeds that are enclosed, protected and nourished within a coating of tissue, variously called a fruit, pod, ovary,

husk or other names. Angiosperm seeds often have special structures to allow dispersal by the wind or by animals, for example, the airborne seeds of dandelions, or the "cling-on" seeds of some *herbaceous* plants. Unquestionable angiosperm remains first appear in the fossil record about 130 million years ago, early in the Cretaceous Period. Fossil evidence indicates that the first angiosperms may have been small herbaceous plants with tiny flowers, or weedy, aquatic plants that had flowers lacking petals and sepals. The Arboretum contains three genera of small herbaceous plants thought to resemble some of the earliest angiosperms: *Houttuynia* (planted in the islands in the parking lot and near the casting area), *Asarum* (found on the north side of the Exhibit Center) and *Saururus* (located near the brook on the west side of the entrance driveway). More than 30 families of flowering plants are represented in the Arboretum; most of these families have a history dating back to the Cretaceous. Currently there are 25 types of magnolias in the collection, 18 types from the witch-hazel family, and eight varieties of the Buxaceae, or "box plant" family. Other notable angiosperm families that likely coexisted with the last of the dinosaurs include katsuras, walnuts, laurels, sycamores, beeches and elms.

Figure 7-4. Conifer trees, such as these reconstructed in the small Triassic-Jurassic diorama in the Exhibit Center, were the dominant plants in the Central Valley during the early Mesozoic. Tough, drought-resistant conifers were well suited to the semi-arid or monsoonal climates that prevailed at the time.

Highlights of the Arboretum

The Arboretum of Evolution at Dinosaur State Park is a horticultural showcase of ancient, exotic, unusual and decorative plants from all over the world. A number of plants in the Arboretum can be seen nowhere else in the State, and the collection is renowned for its diversity and educational importance. A few of the highlights of the Arboretum are briefly described below.

The Fir Collection - The firs are among the most majestic trees in the Arboretum, planted on some of the least disturbed and highest quality soils in the Park (Figure 7-5). Douglas firs are native to the Pacific Northwest, and grow up to 250 feet tall in the wild. They have two rows of soft, flat needles on short stalks. The long seed cones of these firs hang near the ends of branches, and have three-pronged *bracts* extending beyond the cone scales. Nikko firs were introduced to this country from Japan in 1861. This tree has a thick, sturdy trunk and beautiful dark green needles. Nordmann firs, native to the Caucasus Mountains in central Europe, are dark green, spire-like trees; the undersides of their needles are silver. Concolor firs, native to the Rocky Mountains, have long, curved, light-colored, waxy needles up to three inches in length.

The Arborvitae "Wall" - At the northwest side of the Exhibit Center, adjacent to the staff parking lot, more than 50 arborvitae are planted closely together so as to create a dramatic "wall" effect, privacy screen and windbreak (Figure 7-6). Arborvitae are shrub- to tree-size conifers belonging to the cypress family and are often used for landscaping and ornamental purposes. They are valued for their beautiful textures and colors, dense foliage, and variety of sizes and growth habits, though most species are narrow and pyramidal in shape. Arborvitae are native to North America, and some trees can live up to 300 years, hence the name "tree of life." The arborvitae wall contains an array of cultivars, but most notable is the Hiba arborvitae (*Thujopsis dolabrata* 'Variegata'), a species native to Japan (Figure 7-7).

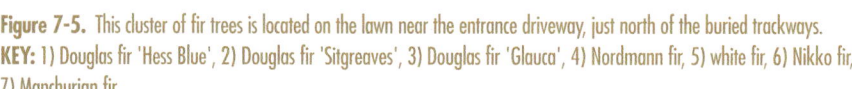

Figure 7-5. This cluster of fir trees is located on the lawn near the entrance driveway, just north of the buried trackways. **KEY:** 1) Douglas fir 'Hess Blue', 2) Douglas fir 'Sitgreaves', 3) Douglas fir 'Glauca', 4) Nordmann fir, 5) white fir, 6) Nikko fir, 7) Manchurian fir.

Figure 7-6. A "wall" of closely spaced arborvitae and a few other evergreen types is planted on the northwest side of the Exhibit Center. The tallest cultivar is 21.5 feet in height.
KEY: 1) 'Emerald Green' arborvitae, 2) western red cedar 'Sunshine', 3) 'Malonyana' arborvitae, 4) 'Ohlendorffii' arborvitae, 5) 'Minima' arborvitae, 6) 'Sudworthii' arborvitae, 7) 'Wansdyke Silver' arborvitae, 8) western red cedar 'Zebrina', 9) 'Lutea Nana' arborvitae, 10) 'Berckman's Golden' oriental arborvitae, 11) 'Juniperoides' oriental arborvitae, 12) Siberian arborvitae, 13) western red cedar 'Collyer's Gold'.

Figure 7-7. Underside of the foliage of the Hiba arborvitae, native to Japan. This evergreen has flattened sprays of glossy, bright green, tiny, scale-like leaves that are a vivid white on the underside. The color contrast between the upper and lower sides of the leaves is striking.

The Cedar Collection - Several drought-resistant cedars (genus *Cedrus*) are planted on the north side of the Exhibit Center in sandy soil that was once backfill adjacent to the short-lived "bubble building" (Figure 7-8). The Lebanon cedar is native to the mountains of the eastern Mediterranean. Many evergreens and even some flowering trees with a strong but pleasant scent are mistakenly called "cedars." The Lebanon cedar is a true *Cedrus*, recognized by its stiff, four-sided needles, which are usually clustered on spurs. Deodar cedars are trees originally from the Himalaya Mountains; in India, this cedar is called the "tree of the Gods." Both the blue-green ('Shalimar') and bright golden ('Aurea') forms of this tree can be seen at the Park. The upright cones of the shalimar look like blue-green eggs perched on the branches. Blue Atlas cedars (*Cedrus atlantica*), native to the Atlas Mountains of North Africa, are wide, irregularly pyramidal trees. The Arboretum contains two types: the 'Glauca', an upright form with silvery-blue needles arranged in whorls on short spurs, and the 'Pendula' form, characterized by dramatic, drooping branches and blue foliage. The California incense cedar (*Calocedrus*), originally from western North America, is an enormous tree in its native habitat, with bright green, aromatic foliage. The wood of this tree is soft and does not splinter easily, and for this reason it is often used for making pencils.

The Pine Grove - An assortment of hardy pine trees is planted on the low ridges bordering the entrance walk and geologic time-line walkway. These conifers do quite well on very compacted subsoils. The collection south of the entrance sidewalk includes: a 'White Mountain' white pine, a silvery version of our native pine with needles in bundles of five; a 'Blue Jay' white pine, a blue-green variety; a Japanese white pine, a small garden pine with tufts of stiff, twisted, blue-green needles; a Himalayan pine, with very long, soft, arching needles; and a Bosnian pine, a small pine that produces deep blue cones in the summer. The collection north of the entrance sidewalk includes two noteworthy pine cultivars: a 'Pendula' white pine, a very full specimen with long, drooping blue-gray needles and branches, and a 'Waterii' Scots pine, which originally ranged from Europe to Siberia (Figure 7-9). This tree is an old, multi-stemmed variety, with pairs of long, steel-blue needles.

Figure 7-8. Part of the cedar collection on the north side of the Exhibit Center.
KEY: 1) Lebanon cedar 'Brevifolia', 2) Lebanon cedar, 3) Deodar cedar 'Shalimar', 4) Blue Atlas cedar, 5) incense cedar, 6) Serbian spruce, 7) Blue Atlas cedar 'Pendula', 8) American yew.

Figure 7-9. Cones of the Scots pine (*Pinus sylvestris* 'Waterii'), on a tree planted on the north side of the entrance sidewalk, near the time-line. The Scots pine is the national tree of Scotland.

The East Side Conifers - A number of interesting conifers is planted on the east and southeast side of the Exhibit Center (Figure 7-10). The Japanese cedar (*Cryptomeria japonica* 'Yoshino') is not a true cedar, but a member of the redwood family. It has awl-shaped needles that are spirally arranged, and reddish bark that peels off in long strips. Its foliage closely resembles that of some of the fossil conifers from the Early Jurassic. The 'Oregon Blue' Lawson cypress is actually a "false cypress" in genus *Chamaecyparis*, and is native to the Pacific Northwest. This tree grows as a narrow, blue column, and has scale-like, feathery foliage in flat sprays. A related form, the weeping Alaskan cedar, also a *Chamaecyparis*, has graceful drooping branches and rich green foliage. These specimens were planted in 1990. The Hinoki false cypress (*Chamaecyparis obtusa* 'Crippsii') is one of the smaller cultivars of this species, but it has bright golden-yellow foliage. The so-called "Chinese fir" (*Cunninghamia lanceolata* 'Glauca') is not a true fir, but rather is a member of the cypress family. This unusual conifer bears stiff, leathery, spirally arranged, lance-like needles like some of the ancient araucaria, fossil relatives of the monkey-puzzle tree. The cultivar term "glauca" refers to the intense blue

Figure 7-10. The plantings on the east side of the Exhibit Center. The highest point on the geodesic dome is 52 feet. **KEY:** 1) weeping Alaskan cedar, 2) weeping Alaskan cedar, 3) Japanese cedar 'Yoshino', 4) wingthorn rose, 5) 'Rheingold' arborvitae, 6) dwarf Arizona fir, 7) 'Berckman's Golden' oriental arborvitae, 8) Leyland cypress, 9) 'Rheingold' arborvitae, 10) Atlantic white cedar 'Heatherbun', 11) Japanese garden juniper, 12) Leyland cypress 'Castlewellan', 13) Lawson cypress 'Oregon Blue', 14) Hinoki false cypress 'Crippsii', 15) Leyland cypress 'Contorta', 16) Leyland cypress 'Silver Dust', 17) Chinese juniper 'Robusta Green', 18) 'Royal Star' magnolia.

color seen on new growth. *Cunninghamia* is very difficult to grow in our area, but at the Park, this tree is in a protected spot near the side of the Exhibit Center.

The Rock Garden - The rock garden, located on a small hill just west of the Exhibit Center, was one of the first areas at the Park to be landscaped and planted as part of the Arboretum (Figure 7-11). Exposed ledges of bedrock are visible in parts of the rock garden, and soils, where present, are thin and dry. Junipers are very tough conifers and can survive in poor, rocky soils. The rock garden features a number of juniper types, some of which grow along the ground in a "creeping" fashion, and others that are taller and more upright. Several junipers are Asian varieties from Japan and Korea. The blue Sargent juniper, native to China, typically grows to 18 inches high, spreads up to 10 feet, and can withstand heat and salt. Small varieties of *Chamaecyparis* ("false cypresses") thrive in the rock garden; these slow-growing evergreens are popular as ornamental plants. Also found are small arborvitae, dwarf firs, and several types of flowering woody or herbaceous plants, including lavender and thyme. Just west of the rock garden, two trees are particularly noteworthy: a Japanese red pine (*Pinus densiflora* 'Umbraculifera') (Figure 7-12) and a golden catalpa (*Catalpa bignoides* 'Aurea').

Figure 7-11. The rock garden is built on bedrock exposures and shallow soils west of the Exhibit Center, near the picnic area. Only very hardy plants can tolerate the poor, rocky soils. **KEY:** 1) Pfitzer juniper 'Aurea', 2) Chinese juniper 'Aurea', 3) Chinese juniper 'Hollywood', 4) Chinese juniper hybrid 'Old Gold', 5) Savin juniper 'Broadmoor', 6) creeping juniper 'Prince of Wales', 7) Chinese juniper 'Keteleeri', 8) Irish juniper 'Hibernica', 9) Chinese juniper 'Spartan', 10) creeping juniper 'Saybrook Gold', 11) Chinese juniper 'Blaauw's', 12) Himalayan juniper 'Meyer's', 13) Atlantic white cedar 'Andelyensis', 14) Chinese juniper 'Robusta Green', 15) threadleaf false cypress 'Filifera', 16) blue Sargent juniper, 17) Hinoki false cypress 'Torulosa', 18) Leyland cypress 'Gold Rider', 19) Lawson cypress 'Golden Showers', 20) Atlantic white cedar 'Glauca Pendula', 21) Japanese garden juniper, 22) common juniper 'Repanda', 23) Chinese juniper 'San Jose', 24) creeping juniper 'Bar Harbor', 25) Savin juniper 'Tamariscifolia', 26) shore juniper 'Blue Pacific', 27) creeping juniper 'Blue Rug', 28) threadleaf false cypress 'Gold Spirals', 29) 'Malonyana' arborvitae, 30) Korean fir.

Figure 7-12. This striking Japanese red pine, planted alongside a pathway just west of the rock garden, is a slow-growing, multi-stemmed cultivar that will attain a height of 20 to 30 feet. **KEY:** 1) dwarf varieties of Norway spruce, 2) hybrid yew 'Hatfieldii', 3) southern catalpa 'Aurea', 4) Japanese red pine 'Umbraculifera', 5) threadleaf false cypress 'Filifera Aurea'.

Figure 7-13. The forecourt at the entrance to the Exhibit Center was nicknamed the "plaza del sol" by Park staff because of its southern exposure and full sunlight most of the year. Benches and attractive plantings make this area a popular gathering place for visitors. **KEY:** 1) Canadian hemlock 'Macrophylla', 2) Japanese white pine 'Adcock's Dwarf', 3) Japanese plum yew 'Ogon', 4) Douglas fir 'Gable's Weeping', 5) Canadian hemlock 'Nana Gracilis', 6) Japanese cedar 'Dacrydioides', 7) Hinoki false cypress 'Nana Gracilis', 8) Hinoki false cypress 'Confucius', 9) Japanese umbrella pine, 10) Japanese cedar 'Yoshino', 11) weeping Alaskan cedar, 12) Hinoki false cypress 'Nana Aurea', 13) Canadian hemlock 'Cloud Prune', 14) Hinoki false cypress 'Pygmaea Aurescens', 15) Himalayan juniper 'Blue Star', 16) Chinese juniper 'Blaauw's Golden', 17) common juniper 'Compressa', 18) shore juniper 'Silver Mist', 19) katsura tree.

The "Plaza del Sol" - The forecourt and entrance to the Exhibit Center face almost due south, receiving much direct sunlight throughout the year. This area was appropriately nicknamed the "plaza del sol" by Park staff. The plaza is a showcase of dwarf and full-size conifers as well as katsura trees and magnolias (Figure 7-13). The plants are arranged to contrast with each other, and they provide much "eye-appeal" by their various forms, textures and colors. The plaza is a popular gathering and resting place for visitors. The Japanese umbrella pines (*Sciadopitys verticillata*) planted here (Figure 7-14) receive more attention than any other conifer in the Park because of their location and highly unusual foliage, which has a "plastic" appearance. The needles of this tree are tiny, flattened scales on the branches. The long, wide, dark green, glossy structures that most people think are needles are actually highly modified branches arranged like the spokes of an umbrella (Figure 7-15). *Sciadopitys* is not a true pine, and is the sole genus in its own family. It has been known in the fossil record for some 230 million years, and thus is another "living fossil" with no known surviving relatives.

Figure 7-14. This pair of Japanese umbrella pines (*Sciadopitys verticillata*) was planted near the entrance to the Exhibit Center in 1991. Fossils of this conifer date back to Triassic times. This tree is not a true pine, and has no close relatives among living plants.

Figure 7-15. Japanese umbrella pines have thick, glossy, dark green branches arranged like the spokes of an umbrella. The unusual foliage has a "plastic" appearance.

The Southeast Collection - The most diverse flora in the United States is found in the Southeastern states. Aspects of the climate of this region are similar to Connecticut's climate in the Early Jurassic. Representative plants from this flora are found in the moist soils that parallel West Street, to the west of the entrance driveway. Among the most interesting trees in this collection is a group of bald cypresses (*Taxodium distichum*), which is planted in a seasonally damp area above buried swamp soils. Bald cypresses are large, majestic conifers with thick trunks at maturity (Figure 7-16). The common name refers to the fact that this *deciduous* tree loses all of its needles prior to winter. The foliage is spirally arranged on the branches; new needles are light green, older needles are dark green, and they turn to a rich brown color in the fall. A pond cypress (*Taxodium ascendens*) is planted nearby. Other trees or shrubs in this collection include American sycamores, a black gum (tupelo), a willow oak, a pair of pawpaws (small trees with distinctive large leaves and fruit), a sweetbay magnolia, a Carolina allspice and a persimmon.

Notable conifers planted in this same area, but which are not part of the Southeast collection, include a giant sequoia, dawn redwoods and a golden

Figure 7-16. These bald cypresses (*Taxodium distichum*) planted near the entrance driveway were started from seeds in 1980. Bald cypresses are native to the Southeastern United States; they thrive at the Park in an area with seasonally moist soil conditions. These conifers lose all of their needles in the winter, hence the appropriate term "bald" in the common name.

larch. The giant sequoia (*Sequoiadendron giganteum*) is recognized as the world's largest tree. In the Sierras of California, its native habitat, this tree can grow to heights of 300 feet. A cultivar of this tree known as 'Hazel Smith' is planted near the large Park sign facing West Street. The 'Hazel Smith' is a hardy variety with short, overlapping, sharp, scale-like needles and distinctive blue-green foliage. The Park has several other sequoia specimens; these are commonly called California redwoods or coast redwoods. The oldest was planted in 1982 near the casting area. Four dawn redwoods (*Metasequoia glyptostroboides*) are planted near the Park sign (Figure 7-17), and like the bald cypresses, these trees lose their needles in the late fall. The dawn redwood was known only from fossils until 1941, when living specimens were discovered in a remote area of China's Szechuan Province. The seeds have since been widely distributed by botanists and cultivators. The short needles are bright green in summer (Figure 7-18), turning to an amber color before they drop. The now massive specimens in the Arboretum were just scrawny *saplings* when replanted here in 1983. The golden larch, another China native, is also deciduous. It has whorls of soft, curved, bright green needles in summer that turn gold and finally orange in the autumn before they are shed. Introduced to this country in 1854, this tree is rarely seen outside of arboreta.

Figure 7-17. Dawn redwoods (*Metasequoia glyptostroboides*) and young spruce trees on the lawn north of the Exhibit Center. Obtained as seedlings in 1979, the now massive redwoods were replanted in this location in 1983. Dawn redwoods are "living fossils" native to China. **KEY:** 1) dawn redwoods, 2) Sakhalin spruce, 3) Wilson's spruce, 4) oriental spruce, 5) Meyer's spruce, 6) Serbian spruce.

Figure 7-18. Dawn redwood foliage. Like the bald cypress, dawn redwoods are deciduous trees, and they lose all of their needles in late autumn.

The Spruce Collection - Several types of spruces are planted on the lawn on the north side of the Exhibit Center (Figure 7-17). The spruce collection grows on moderately fertile, dry soil. Spruces can usually be recognized by their four-sided, spirally arranged needles, which are sometimes sharp-pointed. Their seed cones hang downward. The Hondo spruce is from Japan, and has bold foliage that is shiny green above and bright white underneath. Oriental spruces are slow-growing, densely branched, broadly conical trees superb for small gardens. This tree, native to the Caucasus Mountains, has tiny, light green, glossy needles. The Arboretum contains at least three varieties. Serbian spruces are native to the Balkans; specimens grow into a decorative spire with downward-sweeping, curved branches. The Black Hills spruce, originally from South Dakota, is a slow-growing, dense, pyramid-shaped tree, and is a close relative of our native white spruce.

The East Asian Garden - The plantings alongside the stream east of the entrance drive are exclusively from eastern Asia. Botanists have long recognized that eastern Asia and the Eastern United States share remarkably similar floras. This is especially true of species that are remnants of nearly extinct plant groups. The Asian collection is large and contains several rare or unusual species (Figure 7-19). Among the rarest is a specimen of *Glyptostrobus pensilis* (also called the Chinese swamp cypress or water pine), a conifer similar to the bald cypress, and the only surviving species of its once wide-ranging genus. A weeping katsura tree (*Cercidiphyllum magnificum* 'Pendulum') dominates the small pond at the center of the garden, and an exotic weeping juniper grows nearby. The Asian collection also contains a distinctive "kew" ginkgo, so named because it is a clone of one the first ginkgo trees brought to the West from their native China. The parent tree was planted in 1762 in the Royal Botanic Gardens at Kew, England, where it is still growing. Ginkgoes are true "living fossils" with no

surviving close relatives. A thriving thicket of yellow-grove bamboo is easily noticed from the entrance road. Bamboo is not ancient enough to qualify for membership in the Arboretum of Evolution, but it is a plant highly characteristic of east Asian floras.

The London plane trees lining the entrance drive next to the Asian garden are hybrids of American and Eurasian sycamores, and thus they represent both Eastern and Western floras. These stately trees will eventually form a canopy, creating a living tunnel through which visitors will enter the Park and its unique landscape of ancient plants.

Figure 7-19. Ginkgo (*Ginkgo biloba*) leaves from a tree near the arborvitae "wall." The distinctive leaves are fan-shaped with veins radiating outward from the leaf base. Ginkgoes, native to China, are living remnants of a mostly extinct group of plants. Though no ginkgo fossils have yet been discovered locally, it is very likely that ginkgoes were present in the early Mesozoic forests of the Central Valley.

CHAPTER 8

Natural Areas and Nature Trails

Dinosaur State Park was created to protect and preserve the spectacular Early Jurassic dinosaur trackways at Rocky Hill.

It is important to recognize, however, that the fossil footprints on display are not just stony relics from the past. These tracks also represent living animals successfully surviving in their habitat, coping with short-term and long-term changes in the landscape and climate, and coexisting with other animals and plants in their environment. Visitors to the Exhibit Center can gain much more than an appreciation of dinosaur tracks; the exhibits offer a glimpse into ancient habitats and provide a greater understanding of early Mesozoic geologic history, paleontology, biology and ecology.

Just as the displays in the Exhibit Center illustrate ancient habitats, the natural areas and nature trails at the Park expose visitors to a variety of modern-day habitats and environments.

A broader appreciation of the interactions between modern plants and animals in their habitats sheds light on how early Mesozoic organisms thrived in their environments. Dinosaur Park now maintains more than 2.5 miles of trails within a 60-acre natural area containing a rich diversity of living things in a variety of ecological habitats (Figure 8-1).

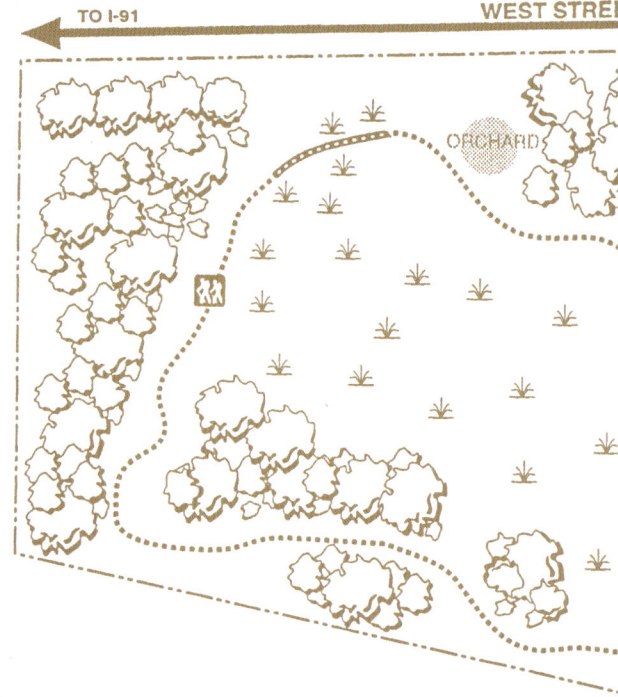

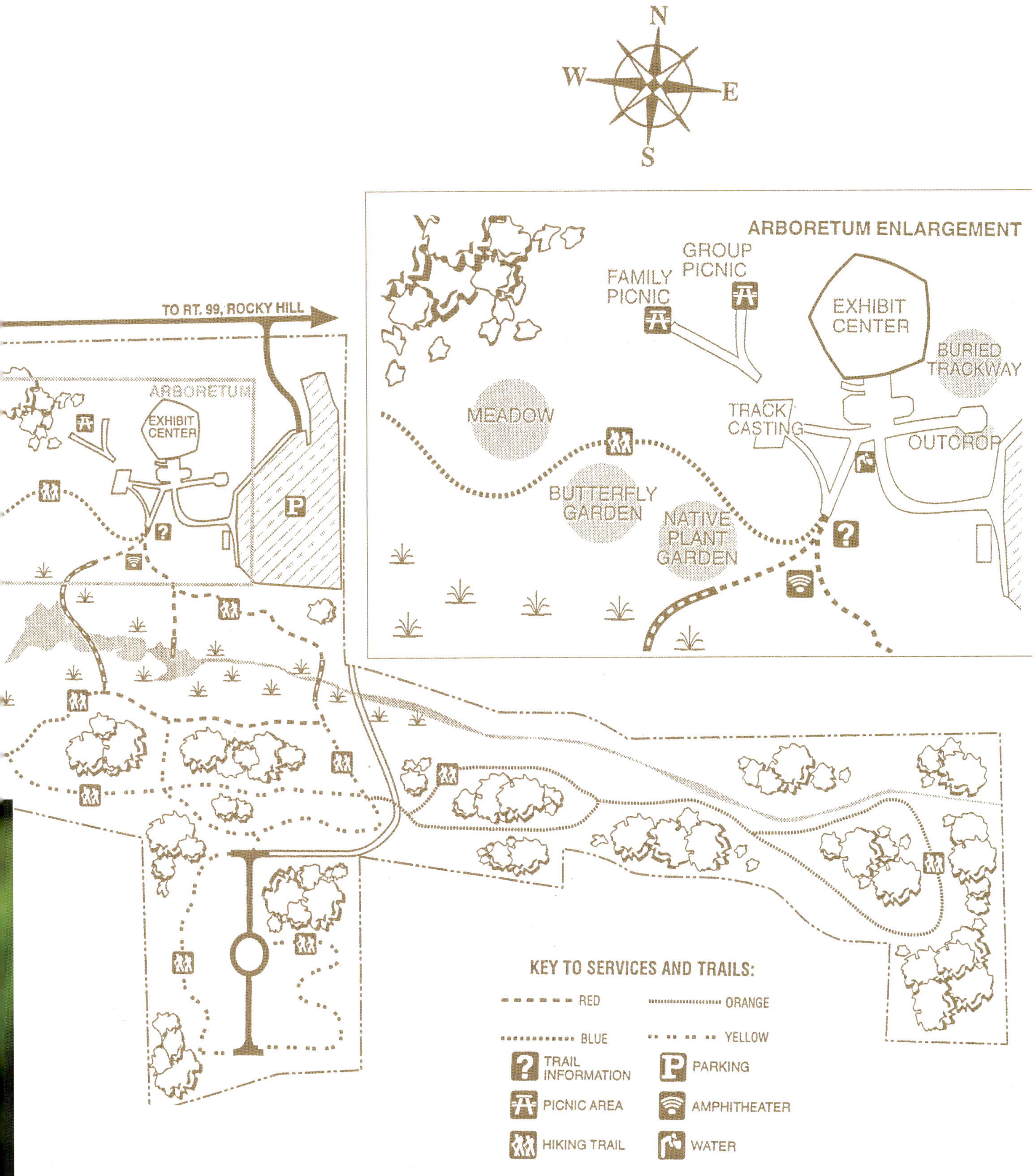

TO RT. 99, ROCKY HILL

ARBORETUM ENLARGEMENT

FAMILY
PICNIC

GROUP
PICNIC

EXHIBIT
CENTER

BURIED
TRACKWAY

MEADOW

TRACK
CASTING

OUTCROP

ARBORETUM

EXHIBIT
CENTER

BUTTERFLY
GARDEN

NATIVE
PLANT
GARDEN

P

KEY TO SERVICES AND TRAILS:

- - - - - RED		·········· ORANGE	
·········· BLUE		· - · - · YELLOW	
？	TRAIL INFORMATION	P	PARKING
🔆	PICNIC AREA	〰	AMPHITHEATER
👫	HIKING TRAIL	💧	WATER

Figure 8-1. Sketch map of Dinosaur State Park, showing the locations of the Exhibit Center and Arboretum, activity areas, and some of the natural habitats found on the Park's 60 acres of woodlands, wetlands and meadows. More than 2.5 miles of color-coded trails pass through a variety of ecological habitats on Park grounds, inviting visitors to explore, discover and appreciate plants and animals in their natural environments. Interpretive signs along the trails describe aspects of Connecticut geology and animal evolution.

The Natural Setting

Perched near the top of a broad bedrock arch known as an *anticline*, the Exhibit Center and trackways at Dinosaur State Park are exactly 200 feet above sea level, slightly higher than much of the surrounding floor of the Central Valley. The highest elevations on Park grounds (some 240 feet) are found on the east-west-trending ridge formed by outcroppings of the Hampden Basalt along the southern boundary of the property. In the center of the Park, sandwiched between the Exhibit Center and the basalt ridge, is a narrow, basin-like valley drained to the east by a small tributary of Hog Brook, which flows into the nearby Connecticut River. Some 15 to 20 acres of the western portion of this "basin" are occupied by densely vegetated wetlands. Except for a small seasonal stream entering the Park from the north and seepage from underground springs, this basin is essentially closed to outside sources of surface water.

The landscape at Dinosaur Park reflects the nature of the underlying Early Jurassic bedrock. The hills and valleys have been molded and shaped by tectonic forces and by weathering and erosion, especially geologically recent glacial activity. All of the bedrock units at the Park, including the Exhibit Center track layers and the basalt ridge, are tilted toward the south, dipping some 11 to 12 degrees from the horizontal. This tilting is a result of faulting and folding that took place during Mesozoic times. The northern continuations of the relatively soft trackway beds were eroded away long ago; the more resistant Hampden Basalt (Figure 8-2) now forms a north-facing ridge and slope. During the most recent ice age, this basalt ridge extended directly across the path of advancing continental glaciers. Continental glaciers tend to smooth off *topography*, eroding high areas and filling in low areas with sediment. Massive glaciers scoured the ridge top, removing loose rock and soil, and plucking basalt blocks from the face of the ridge and carrying them away. The existing bouldery talus slopes visible along parts of the Blue and

Yellow Trails were formed after the glaciers melted away. In the swampy basin between the basalt ridge and the Exhibit Center, posthole excavations have revealed that there is only about a foot of organic muck under the swamp. Beneath the muck there is at least four feet of clean, light yellow, glacially deposited sand containing scattered cobbles. Most of the soils in the Park are of glacial origin; they are generally sandy, with particle sizes ranging from silt to boulders.

Ecological Habitats at Dinosaur State Park

When used in an ecological sense, a "habitat" can be defined as the physical and biological conditions that surround and affect organisms living in a particular area. The natural setting of Dinosaur State Park has created diverse ecological habitats on Park grounds, all within an easy walk from the Exhibit Center. The nature trails at the Park were carefully laid out to pass directly through each of the habitats on the property. Habitats are most easily identified by the plant life they contain. Plants are excellent indicators of ecological conditions because they will thrive only where the environment meets their needs. The major ecological habitats at the Park are listed and briefly described below, with an emphasis on the plant life.

Figure 8-2. Seen in the low-angle light of early winter, volcanic rock of the Hampden Basalt forms a low, east-west ridge along the southern boundary of the Park. Extruded from fissures as an Early Jurassic lava flow, this now tough, durable rock breaks into angular blocks when water seeps into cracks and freezes.

Red Maple and Shrub Swamp – This habitat is found in the lowland area directly south of the Exhibit Center and can be reached via the Red Trail. A 300-foot boardwalk crosses the most interesting part of the swamp (Figure 8-3), where open water provides an ideal location for viewing wildlife. A shorter boardwalk ends at an observation platform in another open area. There is standing water in the swamp except during extended dry periods. Much of the swamp is densely populated by red maples (Figure 8-4), but other trees such as green ashes, willows and American elms also can be seen.

Figure 8-4. Leaf from a red maple (*Acer rubrum*), a fast-growing tree that can tolerate a wide variety of soil and moisture conditions. The name derives from its brilliantly colored autumn foliage and its red flowers in spring.

Figure 8-3. An elevated boardwalk on the Red Trail crosses a swampy lowland where red maples and shrubs are the dominant vegetation. Areas of open water attract a variety of wildlife not seen elsewhere in the Park.

The red maples noticeably diminish in size from larger specimens at the edges of the swamp to thin, scrubby trees on *tussocks* in the center. The maples that live in deeper water grow very slowly and remain small, probably because their roots cannot spread out.

Shrubs also thrive in this perennially wet environment. Common shrubs include sweet pepperbush, shadbush, swamp azalea, swamp rose, highbush blueberries and common alder. The small, open areas are attractive in all seasons, but especially in late spring when the azaleas, roses and pepperbush are blooming and fragrant, and in early autumn, when the maples, blueberries, roses and pepperbush are especially colorful. The wetland habitat also hosts a variety of herbaceous plants such as cattails, pondweeds, smartweeds, duckweed and blue flag iris, as well as abundant low-growing, moisture-loving plants: ferns, sedges, rushes and mosses.

The availability of water and the great variety of plant species in this habitat attract many animals. Diverse populations of insects thrive in and around the swamp, as do amphibians such as newts, salamanders, toads and frogs. The arrival of spring is heralded by a loud chorus of peepers, tiny frogs whose mating calls are one of the earliest signs of the season. Reptiles seen in this habitat include spotted, painted and snapping turtles, as well as garter snakes and common water snakes. Neither of the two poisonous snakes that inhabit Connecticut, copperheads and timber rattlesnakes, has been found on Park grounds. Many mammals visit the swamp and live in the surrounding meadows and woodlands; among the more common are deer mice, wood rats, muskrats, chipmunks, moles and voles, squirrels, bats, raccoons, opossums, weasels, gray and red foxes, *feral* cats, coyotes and deer. Dedicated bird watchers have identified more than 130 species of birds on Park grounds, and wild turkeys are quite common in the area.

Low-Slope Woodlands – This habitat occupies a fairly narrow zone of level to slightly sloping ground surrounding the swamp. The Red Trail passes through much of this habitat. Some of the soils here are acidic and seasonally water-saturated. Swamp white oaks, pin oaks, American elms and red maples are common here, and black gums, ironwood and green and black ash trees can be found, as well as a few gray birches. Several of the swamp shrubs are found in this habitat, and other shrubs include spicebush, nannyberry, whiteberry, partridgeberry and pinxter. Vines such as poison ivy, fox grape, Virginia creeper, greenbrier and carrion flower are more common in this habitat than in the swamp. Compared with other habitats at the Park, the low-slope woodlands contain the greatest variety of ferns, including cinnamon fern, royal fern, marsh fern, New York fern, sensitive fern, crested fern and spinulose wood fern. Mounds of cushion moss and sedges are locally common (Figure 8-5). Patches of horsetails thrive here (Figure 8-6), and these primitive plants look very much the same as their Triassic and Jurassic relatives reconstructed in the dioramas inside the Exhibit Center.

Herbaceous plants do very well in the moist, acidic, organic-rich soils in this area, and under the reduced

Figure 8-5. The dense, fibrous root systems and leaves of tussock sedges form small, thickly vegetated mounds in parts of the swamp and low-slope woodlands.

Figure 8-6. Colonies of horsetails thrive in moist ground at the edges of the swamp. These non-flowering plants reproduce by spores and have hollow, jointed stems. Abundant on land since mid-Paleozoic times, these primitive plants look very similar to their Triassic and Jurassic relatives portrayed in the Exhibit Center dioramas.

light conditions that prevail. Many of the herbs have interesting common names: jewelweed, jack-in-the-pulpit, trout lily, blue flag iris, Canada mayflower, garlic mustard, enchanter's nightshade, beggar's stick-tight, turtlehead, false Solomon's seal, bellwort, false nettle, golden Alexander, and, of course, skunk cabbage. Most of the swamp animals also reside in this area, and other animals pass through this habitat on their way to find drinking water.

Figure 8-7. Mosses are non-flowering plants that grow close together in mats or clumps in shaded or moist areas. These mosses are growing on a dead branch in a shallow pool at the edge of the wet meadow. The brown stalks above the green leaves bear oval spore cases at their tips.

Wet Meadow – This is a small area dominated by grasses at the far western end of the swamp, crossed by a boardwalk on the Blue Trail. Most of the water that enters this habitat is runoff from rainfall and melting snow, but there is also quite a bit of underground seepage, and one spring flows most of the year. In addition to grasses, mosses, sedges and rushes, ferns and horsetails also thrive (Figure 8-7). Red cedar trees can be seen in this habitat, along with shrubs like common alder, multiflora rose, spicebush, silky dogwood, swamp rose, meadowsweet and others. Poison ivy and the *invasive* Asiatic bittersweet are unwelcome vines found here. Other herbaceous plants include boneset, bee balm, dodder, asters, clearweed and

"mad-dog skullcap." In time, this habitat will likely become a woodland, as shrubs and trees, especially red maples, will shade out the grasses and sedges.

Mid-Slope *Mesic* Woodlands – Most of this large, forested habitat lies on the south side of the swamp. The southern loop of the Red Trail borders this habitat, the Yellow Trail passes through it, as do parts of the Blue Trail and much of the Orange Trail. The soils contain moderate amounts of moisture, and the habitat has a cool, north-facing exposure and a gently sloping terrain. These conditions give the mid-slope woodlands a decidedly different character than the low-lying woodlands surrounding the swamp. At the west end of this habitat, black birch trees are dominant, but many of these trees suffer from *cankers* and they are gradually being replaced by longer-lived, shade-tolerant species. The remainder of the habitat is an oak-hickory forest typical of central Connecticut. What is not typical, however, is the large size of many of the oaks (Figure 8-8). The spreading crowns of these oaks are gradually eliminating other trees that need full sun, such as white ashes. American chestnut trees do not grow to full size because of a recurring fungus disease, but they persist as saplings (Figure 8-9). Large numbers of sugar maple seedlings first appeared here in the 1980s, and they now form thickets

Figure 8-8. Former Park director Richard Krueger, a dedicated botanist and avid gardener, stands amidst mature oaks on the Yellow Trail. During his 33-year tenure at the Park, Krueger developed the Arboretum and designed and laid out the trail system that allows access to the diverse natural habitats on Park grounds.

Figure 8-9. An eight-inch-long American chestnut leaf. Once a very common and valuable hardwood tree, American chestnuts now rarely grow to maturity in this country because of a bark fungus that kills the trunk and branches. However, saplings are able to sprout from the roots, and a number of small chestnuts persist in the Park's woodlands. Botanists have been at work for some time to develop disease-resistant chestnuts.

of saplings beneath the older trees in many areas. Additional tree species include black, white and red oaks, bitternut, shagbark and mockernut hickories, American elm, American beech, ironwood, hop hornbeam, black cherry, flowering dogwood and the ever-present red maples. Shrub types include maple-leaf viburnum, low-bush blueberry, arrowwood, huckleberry, alternate-leaved dogwood, partridgeberry and witch hazel. A variety of herbs inhabits the forest floor: spotted pipsissewa, shinleaf, false Solomon's seal, wood and rue anemones, wild geranium, wood aster, wild sarsaparilla, horse balm, violets, rough avens, beechdrops, Pennsylvania sedge and rare red trillium. This is one of the few habitats in the Park that allows the growth of club mosses, such as shining club moss, princess pine and northern running-pine, which are very ancient spore-bearing plants not related to pine trees. Ferns and several types of vines, such as Virginia creeper, poison ivy, dewberry and Japanese honeysuckle, are locally abundant.

Talus Slope - The often steep, bouldery slope of the Hampden Basalt ridge forms a unique habitat along much of the southern boundary of the Park. The Blue Trail and parts of the Yellow Trail follow this narrow band of fragmented rock. This habitat has a rich, herbaceous flora due to the neutral (non-acidic), organic-rich soils that accumulate between the broken basalt blocks (Figure 8-10). The northern exposure and protection from drying summer winds create cool and moist conditions. Trees that thrive here include red, white and black oaks, white ash, black birch, American beech, hop hornbeam, sugar maple and the occasional basswood. Witch hazel, maple-leaf viburnum and bladdernut are shrubs found in this area, and moonseed and Virginia creeper are common vines. Shade-loving ferns often take hold in the moist soils between the boulders; common species are maidenhair fern, marginal woodfern, Christmas

Figure 8-10. Talus slope habitat. The bouldery rubble of the Hampden Basalt forms a cool, moist, north-facing habitat along portions of the Blue and Yellow Trails.

fern, fragile fern, common polypody and broad beech fern. Herbaceous plants abound in this habitat, and the Blue Trail is noted for its spring wildflowers. Among the noteworthy herbs found here are Dutchman's breeches, wild ginger, bloodroot, early meadow rue, red and white baneberries, Solomon's seal, false Solomon's seal, wild columbine, early saxifrage, round-lobed hepatica, rue anemone and rattlesnake plantain.

Ridge-Top and "Old Field" Habitats - The top and south-facing slope of the basalt ridge at the extreme south end of the Park experiences much exposure to sunlight and wind, and only moderately drought-tolerant plants can survive there. The south slope beyond the ridge and areas near the Orange Trail at the east end of the Park were cleared fields and open grasslands in the 1970s. These areas have now become populated by shrubs, fast-growing tree species, and non-native, invasive plant types. At present, these habitats are especially attractive to birds and mammals, providing more food and cover than is available in either forest or field environments. However, these areas eventually will revert back to young forests with a less diverse flora.

Some of the trees that have recently colonized these habitats are bigtooth and quaking aspens, eastern cottonwood, red cedar, crab apples, box elder, black cherry, eastern white pine, green and white ashes, sassafras, flowering dogwood, black and gray birches, and young oaks, red maples and hickories. Shrubs found in these areas include staghorn, winged and scarlet sumacs, pasture juniper, multiflora rose, autumn olive, Tatarian honeysuckle, steeplebush, gray and silky dogwoods, Japanese barberry, European buckthorn, arrowwood, bayberry and winterberry. Vines such as Japanese honeysuckle, Asiatic bittersweet, fox grape, Virginia creeper, catbrier, carrion flower and dewberry contribute to thickets in parts of these habitats. A variety of grasses and other herbaceous plants has replaced ferns as the common ground cover throughout a large portion of these areas. Adding to the sometimes dense undergrowth are herbaceous plants like asters, goldenrods, thistles, cresses, nightshades, buttercups, smartweeds, coneflowers, fleabanes, ragweeds, yarrow, dogbane, black-eyed Susan, St. Johnswort, oxeye daisy, wild carrot, field garlic, evening primrose and burdock.

Grassy Meadow Habitat - This habitat is located west of the casting area and butterfly garden, and can be reached by a short walk on the Blue Trail. Before this land was purchased for the Park, it was part of an apple orchard and later was a commercial nursery planted with yew shrubs. In the early years of the Park, the land became overgrown with brush and ground plants. The area was cleared by Park staff in the 1980s, and was intentionally seeded with native grasses and *forbs* to create a grassland environment (Figure 8-11). Grasslands are an important and endangered habitat in Connecticut, as much open farmland and fields have been converted to industrial or commercial sites or housing developments. The meadow receives full exposure to sunlight, and its soils are dry, sandy and rather poor, but grasses are well adapted for these harsh conditions (Figure 8-12). These grasslands must be maintained by annual mowing to prevent the growth of shrubs and small trees. This environment attracts its own variety of insects, mammals and birds, some of which are not seen in other habitats at the Park (Figure 8-13).

Figure 8-11. Grassy meadow habitat. This area was purposely seeded with native grasses and flowering plants to create a grassland environment. Annual mowing prevents the growth of shrubs and trees. Seen in this view is big bluestem grass, which can grow up to ten feet tall, as well as goldenrod and Queen Anne's lace.

Figure 8-12. Blades of switch grass, heavy with seeds in late summer, arch toward a part of the Blue Trail that passes alongside the meadow.

Among the grasses that can be seen in this habitat are switch grass, Indian grass, big bluestem, little bluestem, sweet vernal grass, redtop grass and orchard grass. Some of these grasses reach heights of ten feet or more. The flowering plants that thrive in this environment include goldenrods, asters, milkweed, vetches, bindweed, speedwells, cinquefoils, Jerusalem artichoke, black-eyed Susan, tick-trefoil, common mallow, bladder campion, bush and red clover, mountain mint, vervain, bull thistle, bluets, smartweeds, yarrow, hawkweeds, buttercups, fleabanes, cresses, wild carrot, monardas, evening primrose, St. Johnswort, mulleins and others.

Figure 8-13. A monarch feeds on nectar from the flowers of an aptly named butterfly bush (*Buddleia*), planted in the Park's butterfly garden near the meadow. Monarchs born in late summer often live for seven months or longer, during which time they make an incredible migration to Mexico, where they spend the winter.

CHAPTER 9

Park Programs

The Exhibit Center at Dinosaur State Park is currently open to the public Tuesday through Sunday, from 9 a.m. to 4:30 p.m. except for Thanksgiving Day, Christmas Day and New Year's Day, when the Park is closed. The entire Exhibit Center is wheelchair accessible. The Arboretum is open daily to visitors during the same hours. The nature trails are also open daily, but they close at 4 p.m. Admission is charged for the Exhibit Center only. A picnic area with tables and grills is available during Park hours (Figure 9-1). Recycling containers are provided for cans and bottles. A brief description of other Park facilities and programs is provided below.

Figure 9-1. A visiting school group enjoys lunch in the shade of the Park's picnic pavilion, located just west of the Exhibit Center.

The Discovery Room - This area within the Exhibit Center features exhibits and interactive activities appropriate for a variety of ages. The room includes exhibits of rocks, minerals and fossils, as well as a greenhouse and a small collection of live animals. Visitors especially enjoy the "hands-on" assortment of dinosaur tracks and other fossils that can be touched and closely examined. There is a children's library collection, as well as puzzles, puppets, and coloring and bookmark-making stations. Scheduled craft activities, demonstrations and presentations by Park staff and invited guests also take place in the Discovery Room.

The Track Casting Area - In this area, located adjacent to the Exhibit Center, Park guests can make a life-size plaster cast of a real dinosaur footprint to take home. Visitors must bring ten pounds of plaster of Paris, 1/4 cup of cooking oil, a large bucket, and rags. Specially designed metal rings provided by the Park are placed around a "negative" *Eubrontes* print. The 200 million-year-old track is then oiled, and after the plaster is mixed with water it is poured over the print (Figure 9-2). The resulting cast is a very realistic "positive" track about a foot long. Track casting is a seasonal activity that is available from May through October.

Figure 9-2. A Park staff member aids young visitors in making a plaster cast from a genuine dinosaur footprint. The cast is molded from a raised negative "out-print" of a *Eubrontes* track from the thick sandstone bed that once lay above the main trackway layers.

The Mining Sluice - At the outdoor mining sluice, visitors sift and search for gemstones and fossils in an old-fashioned, wooden, water-filled sluice, reminiscent of California gold-rush days. There is a small fee for a bag of "mining rough," but participants can keep what they discover. Park staff and volunteers assist visitors in identifying their "finds." This activity enhances observation skills and encourages young people to develop an interest in rock, mineral and fossil collecting. The mining sluice opens in April and closes at the end of October.

Guided Nature Walks - Throughout the summer months, Park staff lead guided walks along the 2.5 miles of nature trails south of the Exhibit Center. The walks focus on geological and biological aspects of the Park's varied natural areas, Connecticut's unique geologic past and the evolutionary history of the animal and plant life typically encountered in the diverse environments. Guided walks begin on weekends during the late spring, and are given daily (except Monday) starting in late June.

Films - On weekends, school vacations and during the summer, the Park offers films for all ages throughout the day in the Exhibit Center auditorium. These include the Park's own original film, *Step Into the Early Jurassic*, which uses maps, animation, and the spectacular trackways, exhibits and dioramas at the Park to reconstruct Connecticut's past during the age of the dinosaurs. Produced as part of the celebration of the Park's 40th anniversary, this 25-minute film features a paleontologist and dinosaur-track expert. The film is available as a DVD in the Bookshop. A wide variety of other films is shown as well.

Guided Programs - From September until June, the Park offers formal guided programs designed to meet Connecticut science standards for each grade level. Programs are available for grades one and up. Teachers and their students receive an interactive slide show, a guided tour of the Exhibit Center and its trackways, and participate in student-centered activities in the Discovery Room. On weekends and in the summer, more informal interpretive programs are held for family audiences. These presentations often highlight the Park's fossil, mineral and rock collections and its resident plants and animals.

Teacher Workshops - The Park sponsors regular workshops for teachers. These daylong sessions are designed to assist teachers in meeting the requirements of Connecticut State Science Standards. Workshops have focused on topics such as Connecticut geology, the fossil record, and on the principles of natural selection and the evolution of life.

Loan Boxes - The Park offers a loan box program for teachers and informal educators. Several loan boxes are available, each one tailored to a different instructional level. The boxes contain books, activities, fossil and mineral specimens, videos, posters and other teaching aids. Educators can borrow these loan boxes for a one-month period for a small deposit. This deposit is refunded as long as the boxes are returned on time with all items in good condition.

Dinosaur State Park Day - This outdoor festival commemorates the discovery of the Early Jurassic trackways and celebrates the Park's enduring value as an educational resource for the general public. Held in late August, this annual event includes games, prizes, live animal demonstrations, crafts, costumed characters (Figure 9-3), nature walks, musical performers, films and track talks in the Exhibit Center. This event attracts well over 2,000 attendees each year.

The Bookshop - The Friends of Dinosaur Park and Arboretum runs the Park's Bookshop. The Bookshop features a wide selection of items including books, toys, custom-designed clothing, amber jewelry, geologic specimens, postcards, dinosaur models, fossils and fossil casts. All profits from the Bookshop support Park programs.

The "Friends" - The Friends of Dinosaur Park and Arboretum, Inc., is a non-profit, volunteer support organization for the Park. Since its establishment in 1976, the FDPA has been an advocate for the protection of the trackways, and has contributed significantly to the Park's educational programs and mission. Resources from membership dues and the Exhibit Center Bookshop are used to fund instructional activities, exhibits, staff and interns, Arboretum plantings and equipment. Major contributions to the Park by the FDPA include the "Triassic" mural by Will Sillin, the Triassic-Jurassic diorama along the east wall of the Exhibit Center, the full-size model of *Dilophosaurus*, and funding for the publication of this volume. Opportunities for volunteers include Bookshop sales, office work, artwork, research, carpentry, gardening, trail maintenance, docent and guide support, and more. The FDPA also publishes the newsletter *Tracks and Trails*, which highlights Park programs and activities and often includes articles of historical or scientific interest.

Figure 9-3. Apparently, not every *Dilophosaurus* was a fearsome predator. This dinosaur, known as "Dilly," routinely appears for Dinosaur State Park Day, and delights children at other special events during the year.

Illustration Credits and Notes

[Photographic illustrations in which a photographer is not credited are by Richard Bergen.]

Figure 1-1. Courtesy NASA.

Figure 1-2. Photo from a slide in Park archives. Photographer unknown.

Figure 2-2. Photo by Andrew Alden, © 2009. Used with permission of About, Inc.

Figure 2-3. Photo by Paul Harrison, March 2005. Used with permission.

Figure 2-6. The original of this specimen is in the collections at the Humboldt Museum of Natural History, Berlin, Germany (HMN 1880). It was collected in 1871. The first *Archaeopteryx* specimen, consisting of a solitary feather, was discovered in 1860; ten specimens now exist in various museums.

Figure 2-7. Courtesy NASA and Continental Dynamics Workshop/NSF.

Figure 3-1. From a figure published in Bell, 1985. Modified by Christine Witkowski.

Figure 3-2. From Christopher R. Scotese, Paleomap Project, 2006. Used with permission.

Figure 3-3. From Christopher R. Scotese, Paleomap Project, 2006. Used with permission.

Figure 3-4. Modified from a diagram by Paul Olsen.

Figure 3-5. These rocks are some of the youngest in the Central Valley and are part of the Portland Formation. Paleobotanist Bruce Cornet provides scale. Photo by the author.

Figure 3-6. Photo by Sigurdur Thorarinsson. Used with permission.

Figure 3-7. Specimen from the East Berlin Formation, in the N.G. McDonald collection.

Figure 3-8. Photo by Matthew Johnson.

Figure 3-9. Specimen from the Shuttle Meadow Formation, in the N.G. McDonald collection.

Figure 3-10. Diagram by Paul Olsen.

Figure 3-11. Diagrams by Paul Olsen.

Figure 3-12. Specimen from the N.G. McDonald collection.

Figure 3-13. Photo by the author.

Figure 4-1. Photo from a slide in Park archives. Original photo from the *Hartford Courant*.

Figure 4-2. Photo from a slide in Park archives. Photo by J.F. Chipps Jr.

Figure 4-3. Photo from a slide in Park archives. Original photo by John Howard.

Figure 4-4. Photo in Park archives by John Howard.

Figure 4-5. Photo in Park archives. Photographer unknown.

Figure 4-6. Photo in Park archives by Sidney Quarrier and Heilpern Photography, modified by Paul Olsen.

Figure 4-7. Photo in Park archives by J.F. Chipps Jr.

Figure 4-8. Photo by Bergen, from an Exhibit Center display photo. Original photographer unknown.

Figure 4-9. Photo by the author.

Figure 4-14. Photo from a slide in Park archives. Original photo by John Howard.

Figure 4-15. Photo from a slide in Park archives. Photographer unknown.

Figure 4-16. Figure published in Farlow and Galton, 2003.

Figure 4-17. Figure published in Farlow and Galton, 2003.

Figure 4-19. Photo by Bergen, from an Exhibit Center display sign.

Figure 5-2. Several of the animal models in the diorama were designed by Donald Baird, formerly a vertebrate paleontologist at Princeton University.

Figure 5-3. Photo by Bergen, modified from an Exhibit Center walkway sign.

Figure 5-8. Sketches by Paul Olsen.

Figure 5-13. In the late 1970s, the Friends of Dinosaur Park Association raised funds for the building of this model, following design details supplied by paleontologist Donald Baird and by dinosaur artist Gregory Paul. The model was fabricated by the Richard Rush Studios of Chicago.

Figure 5-14. This specimen is at the University of California Museum of Paleontology (UCMP 37302), Berkeley, California. Photo from a slide in Park archives. Photographer unknown.

Figure 5-18. Uncatalogued specimen, Division of Vertebrate Paleontology, Peabody Museum, Yale University. Photo by the author.

Figure 5-19. A 1991 photo by the author. Paleontologist Phillip Huber provides scale.

Figure 5-21. Uncatalogued specimens on display in the Joe Webb Peoples Museum, Exley Science Center, Wesleyan University, from the N.G. McDonald collection. Photo by the author.

Figure 5-22. Specimen from the N.G. McDonald collection.

Figure 5-23. Specimen from the N.G. McDonald collection. Photo by the author.

Figure 5-24. Specimen from the N.G. McDonald collection, donated to the Division of Invertebrate Paleontology, Peabody Museum, Yale University (YPM 37598). Photo by the author.

Figure 5-25. Specimens from the Division of Invertebrate Paleontology, Peabody Museum, Yale University. A) YPM 36241. B) and C) YPM 36244. Photos by William Sacco. Figure published in Huber, McDonald and Olsen, 2003.

Figure 5-26. Specimen from the N.G. McDonald collection. Photo by the author.

Figure 5-27. Specimen from the N.G. McDonald collection, donated to the Division of Invertebrate Paleontology, Peabody Museum, Yale University (YPM 37284). Photo by the author.

Figure 5-28. Specimen from the N.G. McDonald collection.

Figure 5-29. Specimen from the N.G. McDonald collection, donated to the Division of Vertebrate Paleontology, Peabody Museum, Yale University (YPM 8605). Photo by the author.

Figure 5-30. Specimen from the N.G. McDonald collection.

Figure 5-31. Figure published in Schaeffer, Dunkle and McDonald, 1975.

Figure 5-32. Specimen from the N.G. McDonald collection.

Figure 5-33. Specimen from the collection of Dr. Alasdair Gilfillan, Bethesda, Maryland.

Figure 5-34. Specimen from the East Berlin Formation, near Durham, Connecticut, in the N.G. McDonald collection. Photo by the author.

Figure 5-36. This locality is on private land and the quarry pits have been filled in. Photo by Bruce Cornet.

Figure 5-37. Specimen from the N.G. McDonald collection.

Figure 6-1. Specimen from the N.G. McDonald collection.

Figure 6-2. Specimen from the N.G. McDonald collection.

Figure 6-3. Photo by the author.

Figure 6-4. Specimen from the N.G. McDonald collection.

Figure 6-5. Specimen from the N.G. McDonald collection.

Figure 6-6. Specimen from Turners Falls, Massachusetts, in the Hitchcock collection, Amherst College Museum of Natural History (AC 27/4). This figure is from Hitchcock's *Ichnology of New England*, 1858.

Figure 6-7. Figure from Hitchcock's *Ichnology of New England*, 1858. The Pliny Moody track slab is on display in the Amherst College Museum of Natural History (AC 16/2). It contains an *Anomoepus* trackway with four distinct footprints, and several other indistinct tracks.

Figure 6-8. Specimen in the Hitchcock collection, Amherst College Museum of Natural History (AC 9/14). Hitchcock (1865, p. 48) suggested that this slab was dug from an old quarry in Middlefield, about two miles west of Middletown, Connecticut. Olsen, Smith and McDonald (1998, p. 589) conclude that the slab more likely was obtained from the quarries in Portland, Connecticut.

Figure 6-9. Diagram by Paul Olsen.

Figure 6-10. Specimen in the Hitchcock collection, Amherst College Museum of Natural History (AC 15/3). Photo by Paul Olsen.

Figure 6-11. Specimen from the collection of Ronald Tavrick, Southbury, Connecticut. Photo by the author.

Figure 6-12. This slab is mounted on the wall of the north lobby of the Exley Science Center, Wesleyan University (WU 185). The tracks have been highlighted with varnish.

Figure 6-13. Specimen from the N.G. McDonald collection.

Figure 6-14. Specimens from a slab mounted on the wall of the south lobby of the Exley Science Center, Wesleyan University (WU 183). The tracks have been highlighted with varnish.

Figure 6-17. Composite drawings from a figure published in Olsen and Rainforth, 2003.

Figure 6-18. Specimen in the Amherst College Museum of Natural History (AC 52/10). This slab bears the type specimen of *Anomoepus curvatus*. Photo by Paul Olsen.

Figure 6-19. Diagram of the trackway of *Batrachopus deweyi*, made from specimens AC 26/5 and AC 26/6 in the Amherst College Museum of Natural History. No locality data is available for these specimens. Figure published in Olsen and Padian, 1986. The largest pes tracks are two inches long.

Figure 6-20. Specimen from the collection of Karl Piela, Chicopee, Massachusetts. Photo by the author.

Figure 6-21. Specimen in the Amherst College Museum of Natural History (AC 41/52).

Figure 6-22. Drawing by Paul Olsen.

Figure 7-1. Diagram by Christine Witkowski.

Figure 7-3. Specimen on display in the Joe Webb Peoples Museum, Exley Science Center, Wesleyan University, from the S. Ward Loper collection (WU 1011). Photo by the author.

Figure 7-4. The small Triassic-Jurassic diorama, located on the east wall of the Dinosaur Park Exhibit Center, was built by Chase Studio of Cedarcreek, Missouri.

Figure 8-1. Courtesy Connecticut Department of Environmental Protection.

Figure 8-3. Photo from a slide in Park archives. Photographer unknown.

Figure 8-8. Photo by Susan Lionberger.

References and Suggested Readings

Alvarez, Walter. 2008. *T. rex* and the Crater of Doom. Princeton University Press, Princeton, NJ, 216 p.

Bell, Michael. 1985. The Face of Connecticut - People, Geology, and the Land. Connecticut Geological and Natural History Survey Bulletin, no. 110, 196 p.

Coombs, W.P. Jr. 1980. Swimming ability of carnivorous dinosaurs. Science, vol. 207, p. 1198-1200.

Cornet, Bruce, Traverse, Alfred and McDonald, N.G. 1973. Fossil spores, pollen, and fishes from Connecticut indicate Early Jurassic age for part of the Newark Group. Science, vol. 182, p. 1243-1247.

Deane, James. 1861. Ichnographs from the Sandstone of Connecticut River. Little, Brown and Company, Boston, MA, 61 p.

Farlow, J.O. and Galton, P.M. 2003. Dinosaur trackways of Dinosaur State Park, Rocky Hill, Connecticut; In: LeTourneau, P.M. and Olsen, P.E., (eds.), The Great Rift Valleys of Pangaea in Eastern North America, volume 2, p. 248-263. Columbia University Press, New York, NY.

Fraser, Nicholas. 2006. Dawn of the Dinosaurs: Life in the Triassic. Indiana University Press, Bloomington and Indianapolis, IN, 307 p.

Hanrahan, Brendan. 2004. Great Day Trips in the Connecticut Valley of the Dinosaurs. Perry Heights Press, Wilton, CT, 253 p.

Hitchcock, Edward. 1858. Ichnology of New England. A Report on the Sandstone of the Connecticut Valley, Especially its Fossil Footmarks. Commonwealth of Massachusetts, Boston, MA, 220 p.

Hitchcock, Edward. 1865. Supplement to the Ichnology of New England. Commonwealth of Massachusetts, Boston, MA, 96 p.

Huber, Phillip, McDonald, N.G. and Olsen, P.E. 2003. Early Jurassic insects from the Newark Supergroup, northeastern United States; In: LeTourneau, P.M. and Olsen, P.E., (eds.), The Great Rift Valleys of Pangaea in Eastern North America, volume 2, p. 206-223. Columbia University Press, New York, NY.

Lockley, Martin and Hunt, A.P. 1995. Dinosaur Tracks and Other Fossil Footprints of the Western United States. Columbia University Press, New York, NY, 338 p.

Lull, R.S. 1953. Triassic Life of the Connecticut Valley (Revised). Connecticut Geological and Natural History Survey Bulletin, no. 81, 336 p.

McDonald, N.G. 1992. Paleontology of the early Mesozoic (Newark Supergroup) rocks of the Connecticut Valley. Northeastern Geology, vol. 14, p. 185-199.

McDonald, N.G. 1995. Connecticut in the age of dinosaurs: a fossil legacy. Rocks and Minerals, vol. 70, p. 412-418.

McDonald, N.G. 1996. The Connecticut Valley in the Age of Dinosaurs: A Guide to the Geologic Literature, 1681-1995. Connecticut Geological and Natural History Survey Bulletin, no. 116, 242 p.

McDonald, N.G. and LeTourneau, P.M. 1988. Paleoenvironmental reconstruction of a fluvial-deltaic-lacustrine sequence, Lower Jurassic Portland Formation, Suffield, Connecticut; In: Froelich, A.J. and Robinson, G.R. Jr., (eds.), Studies of the Early Mesozoic Basins of the Eastern United States. U.S. Geological Survey Bulletin, no. 1776, p. 24-30.

McHone, J.G. 2004. Connecticut in the Mesozoic World. Connecticut Geological and Natural History Survey, Special Publications, no. 1, 40 p.

Milner, A.R.C. and Kirkland, J.I. 2007. The case for fishing dinosaurs at the St. George [Utah] dinosaur discovery site at Johnson Farm. Survey Notes (Utah Geological Survey), vol. 39, no. 3, p. 1-3.

Olsen, P.E. and Padian, Kevin. 1986. Earliest records of *Batrachopus* from the southwestern United States, and a revision of some early Mesozoic crocodylomorph ichnogenera; In: Padian, Kevin, (ed.), The Beginning of the Age of Dinosaurs: Faunal Change Across the Triassic-Jurassic Boundary, p. 259-273. Cambridge University Press, Cambridge, England.

Olsen, P.E. and Rainforth, E.C. 2003. The Early Jurassic ornithischian dinosaurian ichnogenus *Anomoepus*; In: LeTourneau, P.M. and Olsen, P.E., (eds.), The Great Rift Valleys of Pangaea in Eastern North America, volume 2, p. 314-368. Columbia University Press, New York, NY.

Olsen, P.E., Smith, J.B. and McDonald, N.G. 1998. Type material of the type species of the classic theropod footprint genera *Eubrontes*, *Anchisauripus*, and *Grallator* (Early Jurassic, Hartford and Deerfield basins, Connecticut and Massachusetts, U.S.A.). Journal of Vertebrate Paleontology, vol. 18, p. 586-601.

Ostrom, J.H. 1972. Were some dinosaurs gregarious? Palaeogeography, Palaeoclimatology, Palaeoecology, vol. 11, p. 287-301.

Rainforth, E.C. 2003. Revision and re-evaluation of the Early Jurassic dinosaurian ichnogenus *Otozoum*. Palaeontology, vol. 46, p. 803-838.

Schaeffer, B., Dunkle, D.H. and McDonald, N.G. 1975. *Ptycholepis marshi* Newberry, a chondrostean fish from the Newark Group of eastern North America. Fieldiana: Geology, vol. 33, p. 205-233.

Schaeffer, B. and McDonald, N.G. 1978. Redfieldiid fishes from the Triassic-Liassic Newark Supergroup of eastern North America. Bulletin of the American Museum of Natural History, vol. 159, p. 129-174.

Weishampel, D.B. and Young, Luther. 1996. Dinosaurs of the East Coast. Johns Hopkins University Press, Baltimore, MD, 275 p.

Welles, S.P. 1954. A new Jurassic dinosaur from the Kayenta Formation of Arizona. Bulletin of the Geological Society of America, vol. 65, p. 591-598.

Zeilinga de Boer, Jelle. 2009. Stories in Stone. How Geology Influenced Connecticut History and Culture. Wesleyan University Press, Middletown, CT, 206 p.

Glossary

Glossary terms in the text are in italics the first time they appear; most capitalized *and* italicized words in the text are genus/species names, and are not in this glossary.

alluvial fan A gradually sloping deposit of sediment that widens out like a fan, formed where a swift-moving stream abruptly slows down, as when an upland stream emerges onto a level plain at the foot of a mountain range.

amphibians Scaleless, backboned animals that usually emerge from eggs laid in water as gill-bearing tadpoles, and later develop lungs. No amphibian has claws. Salamanders and frogs are examples.

angiosperms Plants that produce flowers and have seeds enclosed in a fruit, pod or other structure.

anoxic Completely deprived of oxygen.

anticline An arched fold in which the rock layers slope downward, usually in opposite directions, away from the central axis of the fold. The oldest rocks are in the core of the fold.

arthropods Invertebrate animals with a segmented body and jointed legs, such as insects and spiders. Fossil evidence suggests that arthropods were the first animals to colonize land.

artifacts Objects made by humans, including simple tools, weapons, pottery and the like.

basalt A hard, dark-colored igneous rock formed from low-silica lava; weathers to a rusty color.

bipedal Refers to an animal that walks on two feet.

brachiopods A once-widespread group of invertebrate animals with hinged shells enclosing tentacles used for guiding food particles toward the mouth.

bracts Modified leaves that are usually small and scale-like.

braided river A stream that has a network of branching channels separated by islands and bars.

brownstone A reddish-brown sandstone common in the Central Valley; very popular as a building stone in the 19th century.

cankers Plant diseases that cause ulcer-like decay of bark and wood.

Central Valley The broad, elongate lowland region of central Connecticut and Massachusetts containing rocks of Triassic or Jurassic age. The Connecticut River flows through much of the Central Valley, but the Valley is much older than the River.

club mosses Primitive, flowerless plants with a moss-like appearance, typically having club-shaped, spore-bearing cones at the end of stems or short branches.

coelacanths A group of primitive fishes best characterized by large, fleshy, lobed fins. Coelacanths likely were the direct ancestors of land vertebrates. In ancient times they inhabited freshwater environments as well as the sea.

conglomerate A coarse-grained sedimentary rock formed from the cementation of rounded pebbles or gravel.

conifers Cone-bearing trees and shrubs, such as pines, spruces, firs, etc. Conifers were the dominant land plants for much of the Mesozoic.

Cretaceous Period The youngest of the three periods in the Mesozoic Era, beginning about 145 million years ago. The Cretaceous follows the Triassic and Jurassic Periods. The name comes from "creta," the Latin word for chalk (white limestone), and was originally applied to certain rocks in the Paris Basin, France.

crustaceans Arthropods with a hard outer skeleton, such as shrimps, crabs and lobsters.

cultivars Plant varieties or types developed by humans.

cycadeoids Extinct cycad-like plants typically possessing a short, thick trunk and a crown of leathery, fern-like leaves.

deciduous Refers to plants that shed their leaves annually or during certain seasons; the opposite of "evergreen."

diabase An igneous rock similar to basalt, formed from magma that solidifies at shallow depths in the crust.

dip The angle a rock unit or geologic feature makes with a horizontal plane.

ecology The branch of biology that studies the relationships between organisms and their environment.

ecosystems Communities of plants, animals and other organisms interacting with each other and with their chemical and physical environment.

eons The largest units of the geologic time scale, sometimes lasting billions of years.

eras Subdivisions of eons in the geologic time scale, containing two or more periods.

faults Fractures in bedrock along which there has been movement of the sides relative to each other.

feral A term referring to a wild animal or a domesticated animal that has become wild.

fishes The plural of fish when referring to individuals of different species.

fissures Long, narrow, deep cracks in the Earth's crust, often created by spreading plates.

forbs Broad-leaved flowering plants, as distinguished from grasses, sedges, etc.

formations Bodies of rock of sufficient thickness and character so as to be easily distinguished from other rock units.

fossils Traces of past life, usually preserved in rock. For convenience, many geologists agree that fossils must be at least 10,000 years old. Traces of life younger than 10,000 years are called "remains."

genera The plural of genus.

genus A level of biological classification between family and species. Closely related species often are included in the same genus. Dogs, coyotes and wolves, for example, are all in the genus *Canis*. The genus name is the first word in "binomial nomenclature" — the two-word scientific name for an organism.

geologic time The vast interval of time from the formation of the Earth as a planet to the present day. Many geologic processes take place over such great spans of time that most "human" time measurements (minutes,

hours, days) are meaningless. A million years is a relatively "short" time in geologic history.

habitats Areas where an organism naturally grows or lives; its native environment.

hallux The innermost toe of the foot, equivalent to the "big toe" on a human foot.

herbaceous Describes a seed plant whose stem withers away after each growing season; the opposite of a woody-stemmed plant.

herbivores Animals that feed on plants.

hominids All forms of extinct and living humans.

horsetails Primitive spore-bearing plants with hollow, jointed stems; scale-like leaves emerge from the stem joints and the spores are contained in a large cone-like structure at the tip of the stem. The name comes from the resemblance of some types to a horse's tail.

ichnogenera Genus names applied to a footprint or other trace fossil.

ichnospecies Species names applied to a footprint or other trace fossil.

in situ A term referring to something found in its original place; not moved.

invasive Describes non-native plant or animal species that heavily colonize a habitat, often with harmful effects to native species or that habitat.

invertebrate Any animal without a backbone or similar structure.

Jurassic Period The middle of the three periods in the Mesozoic Era, beginning about 200 million years ago. The Jurassic is named for the Jura Mountains, at the borders of Germany, France and Switzerland, in which rocks of this age were first studied.

lacustrine Relating to a lake or lakes.

mammals Warm-blooded, vertebrate animals whose offspring are nourished by milk-secreting glands.

manus Latin for "hand." Generally refers to the forefoot of a four-limbed animal.

mesic Moderately moist; a term typically applied to plants or to a habitat.

microlaminated A term used to refer to the very thin light and dark layers seen in some shales and other rocks. Typical microlaminations are usually less than one millimeter thick.

mountain belts A series of connected mountain ranges and chains formed by converging plates; often aligned parallel to the coastlines of continents. The Andes and Appalachian Mountains are examples.

muscovite A colorless to pale-brown mineral of the mica group that readily splits into flat sheets or flakes; common in many different types of rocks.

niche The particular role or ecological space that an organism occupies within its habitat; how an organism makes a living in its environment.

oncolites Concentrically laminated, typically rounded, calcareous masses made by blue-green algae (also called cyanobacteria) in shallow seas and lakes.

ornithischians An extinct group of beaked, herbivorous dinosaurs, named for their bird-like hip structure. Birds, however, are the descendants of saurischians, the "lizard-hipped" dinosaurs.

paleontologists Scientists who specialize in the study of fossils.

Pangaea A "supercontinent" that broke apart about 200 million years ago to form the present configuration of the continents.

penetrative tracks Tracks typically made when an animal steps deeply into very soft, muddy sediment. Penetrative tracks document the behavior of the sediment in response to the entrance and exit of the animal's feet, but they do not often provide an accurate record of the shape of those feet.

periods Subdivisions of eras in the geologic time scale.

pes Latin for "foot." Generally refers to the rear foot of a four-limbed animal.

plates Large, thick, mobile slabs of the Earth's crust and upper mantle rock.

plate tectonics The concept that the Earth's surface is composed of massive interacting plates that are slowly moving and changing in size and shape. Intense geologic activity (folding, faulting, earthquakes and volcanism) often takes place at the boundaries of plates.

playas Flat-floored basins in arid regions that temporarily become shallow lakes after heavy rains.

primates A group of mammals, including monkeys, apes and humans, that have large brains, heightened vision, and hands and feet with five flexible digits.

prosauropods A diverse group of herbivorous dinosaurs that lived during the Triassic and Jurassic Periods. Characterized by a long neck and tail and a small head, these reptiles had forelimbs that were shorter than their hind limbs. The manus was equipped with a very large thumb claw. Some prosauropods reached lengths of 30 feet.

quadrupedal Refers to an animal that walks on four feet.

remains Traces of past life less than 10,000 years old. Remains older than 10,000 years are called "fossils" or "fossil remains."

reptiles Traditionally defined as cold-blooded, backboned animals with lungs and a body covering of scales or bony plates, reptiles include snakes, lizards, turtles, crocodiles, etc. Dinosaurs are considered to be reptiles, though some of them had feathers and probably were warm-blooded. Birds, likely descended from theropod dinosaurs, are now also classified as reptiles by most vertebrate paleontologists.

rhizomes Typically horizontal plant stems lying atop or under the soil, such as in ferns or grasses.

rifting The thinning, splitting and spreading apart of plates due to rising currents of hot mantle rock.

rills Small networks of narrow, shallow channels caused by flowing water.

saplings Young trees.

shale A fine-grained sedimentary rock formed by the cementation of silt and clay (mud). Shale readily splits into thin layers.

species As traditionally defined, a species consists of very closely related organisms that can interbreed and produce fertile offspring under natural conditions; a distinct form of life. The species name is the second word in "binomial nomenclature" — the two-word scientific name for an organism.

strata Layers of sedimentary rock.

talus slope An inclined pile of rock debris at the base of a cliff.

tetradactyl Having four digits (fingers or toes) on a limb.

theropod Any of a diverse group of typically carnivorous bipedal dinosaurs. Theropods first appeared in the Late Triassic. The well-known *Tyrannosaurus rex* was a Late Cretaceous theropod. Today, theropods are represented by birds, their living descendants.

topography The shape, size and position of natural or man-made surface features of the landscape, including hills, valleys, rivers, etc.

trace fossils Sedimentary structures that display evidence of the life activities of an organism, such as tracks, trails, burrows, etc., but do not preserve any of the actual body parts of that organism.

Triassic Period The oldest of the three periods in the Mesozoic Era, beginning about 250 million years ago. The Triassic was originally named for three distinct rock types of this age in Germany.

tridactyl Having three digits (fingers or toes) on a limb.

trilobites A diverse but long-extinct group of marine arthropods that thrived during the Paleozoic Era. The trilobite body was covered by an external skeleton divided into three regions. Trilobites slightly resembled the modern horseshoe crab.

tubercles Small, rounded projections or growths.

tubers Short, thickened, fleshy parts of an underground stem; potatoes are an example.

tussocks Thick tufts or clumps of grass or sedge, often bound together by the roots.

type specimen A single biological or fossil specimen upon which the original description and scientific name of that organism is based.

underprints Vague, shallow track impressions made in sediment layers beneath the actual surface layer the animal walked upon.

vascular plants Plants that have specialized tissues (xylem and phloem) that conduct water, minerals and photosynthetic products such as sugars.

vestigial A term referring to an organ or structure in a species that has no known function. The appendix in humans is an example.

whorl A circular growth of leaves, petals, etc., originating from the same point on a stem.

zooplankton The mostly microscopic animal life found in the ocean or in bodies of fresh water.

Index

[Index selections found in figure captions are denoted by an *f* following the page number.]

About The Author

Nicholas G. McDonald is a geology and biology instructor at Westminster School in Simsbury, Connecticut, and is emeritus Chairman of the Science Department. His abiding interest in paleontology was awakened at the age of 12 during a Sunday afternoon field trip to a local Jurassic fossil site. After graduating from Franklin and Marshall College, he continued his geologic studies at Wesleyan University, completing a Master's thesis on the fossil fishes of Connecticut and Massachusetts. An inveterate collector, Nick has brought to light literally thousands of well-preserved specimens from early Mesozoic rocks once thought to be largely devoid of fossils. His discoveries of fishes, plants, mollusks, insects, crustaceans and trace fossils have broadened the understanding of Jurassic ecosystems in the region. His publications include a number of scholarly papers and a bibliographic volume: *The Connecticut Valley in the Age of Dinosaurs*. For more than 20 years the author has been a Visiting Scholar (research associate) in the Department of Earth and Environmental Sciences at Wesleyan, and he is currently a Curatorial Affiliate at the Yale Peabody Museum of Natural History.